On-Site Stormwater Management

Applications for Landscape and Engineering
Second Edition

On-Site Stormwater Management

Applications for Landscape and Engineering
Second Edition

Bruce K. Ferguson
School of Environmental Design, University of Georgia

Thomas N. Debo
City Planning Program, Georgia Institute of Technology

Library of Congress Catalog Card Number 89-77729
ISBN 0-442-00116-9

Printed in the United States of America

Van Nostrand Reinhold
115 Fifth Avenue
New York, New York 10003

Van Nostrand Reinhold International Company Limited
11 New Fetter Lane
London EC4P 4EE, England

Van Nostrand Reinhold
480 La Trobe Street
Melbourne, Victoria 3000, Australia

Nelson Canada
1120 Birchmount Road
Scarborough, Ontario M1K 5G4, Canada

16 15 14 13 12 11 10 9 8 7 6 5 4 3 2 1

Library of Congress Cataloging-in-Publication Data

Ferguson, Bruce K.
On site stormwater management: applications for landscape and engineering / Bruce K. Ferguson. Thomas N. Debo.—2nd ed.
p. cm.
Includes bibliographical references.
ISBN 0-442-00116-9
1. Urban hydrology. 2. Watershed management. 3. Urban runoff. I. Debo, Thomas N. (Thomas Neil), 1941- II. Title.
TC409.F47 1990
628'.21—dc20 89-77729
CIP

Contents

	Preface	Pages 1 - 2
Chapter 1	**Introduction**	Pages 3 - 18
Chapter 2	**Storm Runoff Estimation**	Pages 19 - 54
Chapter 3	**Water Balance Estimation**	Pages 55 - 78
Chapter 4	**Conveyance Design**	Pages 79 - 104
Chapter 5	**Storm Detention Design**	Pages 105 - 120
Chapter 6	**Extended Detention Design**	Pages 121 - 132
Chapter 7	**Infiltration Design**	Pages 133 - 152
Chapter 8	**Water Harvesting Design**	Pages 153 - 174
Chapter 9	**Epilogue**	Pages 175 - 176
	Bibliography	Pages 177 - 184
	Acknowledgments	Pages 185 - 186
Appendix A	**Exercise Sites**	Pages 187 - 200
Appendix B	**Drainage Area Analysis**	Pages 201 - 210
Appendix C	**Grading of Basins**	Pages 211 - 214
Appendix D	**Sediment in Basins**	Pages 215 - 218
Appendix E	**SCS Time of Concentration**	Pages 219 - 226
Appendix F	**Open Ephemeral Basins**	Pages 227 - 230
Appendix G	**Safety and Liability**	Pages 231 - 234
Appendix H	**Software Operation**	Pages 235 - 256
Appendix I	**Constants and Conversions**	Pages 257 - 260
	Index	Pages 261 - 270

On-Site Stormwater Management

Applications for Landscape and Engineering
Second Edition

Preface

The subject of stormwater management is a specialized but increasingly important area of landscape architectural and site engineering practice.

So this book is intended to supplement broad basic training in grading, drainage and site construction with basic concepts of stormwater hydrology and their implications for land use, urban design and the environment.

Its major purposes are:

1. Recognition of the range of *qualitatively* different management methods and site conditions, and their implications for water resources and potential site development.
2. Familiarity with basic concepts of *quantitative* on-site stormwater hydrology and approaches used to estimate and design for runoff.

The level of quantitative hydrology covered in this book is basic. Further sophistication in hydrologic theory or modeling requires intensive specialized courses, resident graduate training or extensive specialized experience.

The material in this book was originally developed for continuing education courses for practitioners. For several years the material was in the form of informal handouts for discussion, temporary use and gradual upgrading. The intense interest of practitioners around the country in stormwater in general, and our courses in particular, forced us to make a commitment to formal publication. The first edition of this book proved itself as a useful reference for practitioners. It was also found appropriate in university classes, since it pointed out the kind of work that students would be asked to do when they, too, entered practice.

This second edition represents a major rewriting of the book. Every page has been revised to reflect evolution in the state of the art. New technologies have been born, older technologies have been reevaluated, and "basics" are being performed in different ways. We retain hope that the formal nature of this book will not inhibit future upgrading in response to further evolution in the science of hydrology and the art of design.

Chapter 1

Introduction

• **A young designer** in a multi-discipline consulting firm is gradually being given all the responsibilities connected with site development projects, from master planning through engineering design and construction inspection. To round out the techniques that he is responsible for carrying out, he needs to enlarge his skills in drainage design and stormwater control.

• **A project designer** in a prestigious design office needs in her daily work to integrate complex systems of water features and land use. She is asking fundamental questions about the presence and movement of water, aimed at making site amenity and environmental management work together.

• **Two designers** in the same multi-discipline office work regularly on residential and commercial developments in their area. Local ordinances regularly require them to complete stormwater control plans for their projects. Recently the standards and procedures in those ordinances have been changing.

• **The head** of a small design office is not being required by either her clients or local ordinances to control stormwater, but she believes that any developer is obligated to control the impacts of his development, and she wants to convince her clients to adopt her viewpoint.

• **A subdivision review officer** in a planning commission is dismayed at the poor aesthetic, safety and land use aspects of the stormwater facilities being proposed by developers in her rapidly growing area. To encourage people to do better, she has to educate herself in what the alternatives for stormwater management are and how to determine which ones are feasible on each development site.

• **The chief designer** in a large company that builds recreational vehicle parks is expected by his employer to be aware of and to respond to all the technical and legal ramifications of his designs. Stormwater is one of the most visible and controversial of those ramifications.

These are some of the people who have come to our short courses on stormwater management in recent years. They are all on-the-line practitioners with substantial responsibility for their projects. They are either being forced to respond to pressures to control stormwater properly, or they are self-motivated to bring the latest in design standards and technical competence to their work.

In today's world, people who develop land have to control their stormwater runoff. All site developments, of all kinds, involve impervious surfaces which increase runoff over the surface, dump flood waters into streams, reduce groundwater recharge, divert water from stream base flows, and turn oil and rubber from the streets into pollutants. Our society demands that water quality must be protected, floods suppressed, droughts prevented, water conserved. These demands are reflected partly in courts of law, where people whose property is damaged by a development's uncontrolled storm water have a right to be compensated. More directly, the demands are reflected in specific stormwater and water quality regulations.

Also in today's world, high standards are set for the amenity, safety, economy and resourcefulness of site developments. For a channel, a basin, or a lake, alternative constructions must be examined and multiple concerns accommodated. Facilities for controlling storm water must be smoothly integrated into the land use and activity patterns of their communities.

Urban communities need safety of life and property from floods, flowing streams for amenity and the environment, and secure quality and quantity of water supplies. It is not enough to dispose surface waters into streams during storms. On-site storm water must be managed.

This book is aimed at people who have a job to do in a world characterized by growing scientific understanding, developing technologies and environmental issues. Its emphasis is on practical applications, not background explanations.

It is not meant to make you a stormwater know-it-all. It does not cover aspects of the very large field of hydrology other than those applying directly to on-site stormwater. Nor does it cover the physical materials of stormwater facilities such as dams, weirs or channels. Carrying out a hydrologic program through physical layout and construction remains a matter for creative urban design, beyond the field of hydrology.

This book does give you the immediate ability to do specific practical applications. There are *qualitative* choices to be made in the types of water movements we want to create, and the types of facilities they involve. There are also *quantitative* calculations for rates of flow and the sizes and shapes of facilities to bring intended functions about. Making the qualitative choice — determining what kind of hydrologic function or pattern a site will have — is possibly the most important step in stormwater control. Once the qualitative choice is made, quantitative models can be used to define the magnitudes of flows and storages within that choice.

For small, simple, on-site systems the level of mathematics illustrated in this book should enable you to do all the calculations necessary to complete a hydrologic program. The formulas given are basic and do not cover all possible site-specific circumstances. For further mathematical details designers can refer to sources listed in our bibliography.

For master planning large projects such as community plans or large public works this book covers the reasoning behind management options. It enables you to understand the objectives and constraints that hydrologic concerns may impose upon a master plan. With the basic concepts introduced in this book you can communicate with specialists doing hydrologic details, although they may be using quantitative models much larger and more complex than those illustrated here.

The responsibility for applying any method to a specific site, and the liability for its results, rests with the licensed professional in charge of the project, not with the authors or publisher of this book. Accepted regulations and practices vary from region to region and office to office. Specific needs vary from site to site. This book covers enough of the basic concepts of stormwater hydrology to enable you to learn a range of calculation methods. Local standards and practices should be used for site-specific design.

At its roots hydrology is fundamentally simple. Since we will be talking about hydrology, we want you to be comfortable with the basic physical concepts that underlie it. It amounts merely to moving volumes around over time. It is underlain by — indeed, almost entirely composed of — the basic concepts of volume and rate that we all learned in elementary school. So if you are comfortable with those basic definitions, you can be comfortable with all the ways we will look at water in this book. To show you how simple the foundations of hydrology are, let us review everything that most people know about volumes of water.

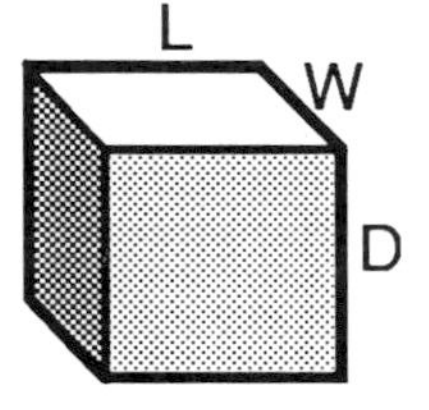

A static volume such as that of a cube is equal to the product of the three sides,

$$\text{Volume} = LWD$$

We all learned that in elementary school. Since volume is the product of three dimensions, its units are cubic, such as cubic feet (cu.ft.). Since LW equals area, we can also say that volume is equal to the product of area and depth,

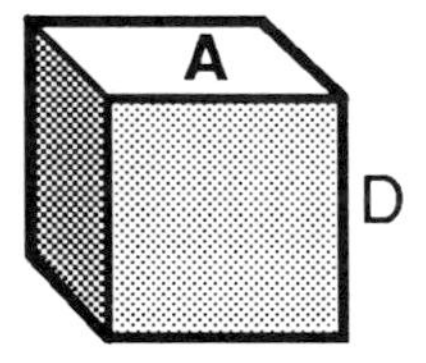

$$\text{Volume} = AD$$

Thus we could analyze a volume of water from the side (depth), from the top (area) or as a total volume.

Velocity is defined as distance per time,

$$\text{Velocity} = \text{Distance/Time}$$

So we use corresponding units such as feet per second (fps). This might be velocity of flow through a pipe, or velocity of a ponded water surface approaching the soil as the water infiltrates.

Rate of flow comes from combining the concepts of volume and velocity. If we dump a bucket of water on the road, we could say that we have created a flow of one cubic foot during the time we were there. If we keep on dumping more buckets, all those volumes of water flowing down the gutter add up to a stream of volumes over time. We could express the stream's magnitude in units such as cubic feet per second (cfs). Any culvert, any swale, any river, any water supply pipe has a flow which we can express the same way.

Any rate embodying the same concept of volume per time is symbolized by Q:

$$Q = \text{Rate of flow} = \text{Volume/Time}$$

Since volume = (area)(depth), and velocity (V) = distance or depth per time, we can also say,

$$Q = (\text{Area})(\text{Velocity}) = AV$$

The above equation can show us velocity when we know the rate of flow and the cross-sectional area through which water is flowing:

$$V = Q/A$$

Typical units in the above equation would be fps for velocity, cfs for flow and sq.ft. for area.

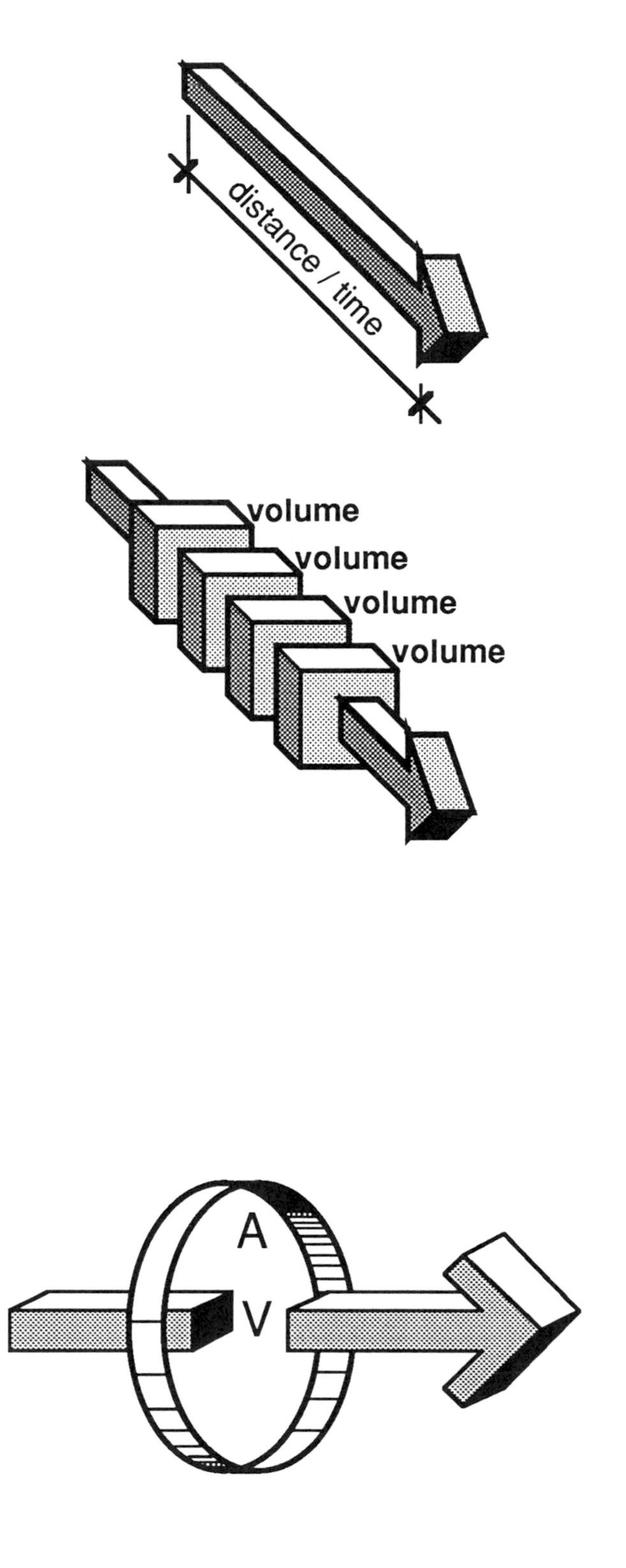

Total volume of flow results when a rate of flow has gone on for a period of time. By simple algebra, volume of flow (*Qvol*) is the product of the rate of flow (volume/time) and the time over which the flow continues:

$$Qvol = \text{Volume of flow} = (Q)(\text{time})$$

Volume of flow is expressed in cubic units such as cubic feet or acre-feet, since we have ended up back at a static volume.

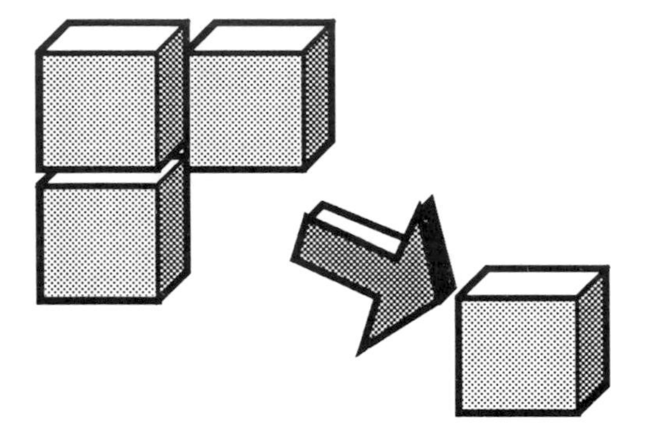

The change in volume in a standing body of water such as a stream, a reservoir or your bathtub is simply the difference between inflow and outflow over a period of time:

$$\Delta\text{Volume} = Qvol_{\text{in}} - Qvol_{\text{out}}$$

It is still expressed in static volumetric units such as acre feet. The water level in a water body fluctuates as its volume rises and falls.

A hydrograph is a useful summary of stormwater flows. It is a plot or table of rate of flow (*Q*) against time (*T*). The figure shows a hydrograph of runoff from a small watershed during a theoretically typical storm event. The bottom (horizontal) scale is time. At the side (the vertical scale) is *Q* in units such as cfs.

***Q* varies** during the time of the storm. At the beginning runoff (*Q*) is very low. As the storm progresses runoff builds to a peak. The rate of flow at the peak moment is designated *Qp,* the time at which it occurs *Tp.* As the storm recedes the rate of runoff gradually declines to its original low level.

A hydrograph also shows *Qvol.* Imagine that we put the hydrograph through a bread slicer, cutting it into many thin vertical slices. Each slice approximates a tall, skinny rectangle. Each slice is a few minutes in width. The rate of flow during that moment is shown by the height of the curve at that slice. The *Qvol* in each slice is *Q* multiplied by time, which is the area of the rectangle (width times height).

***Qvol* for the storm as a whole** is the sum of the *Qvol*s in all the slices, which is the area under the whole runoff curve: *Q* extended over a period of time.

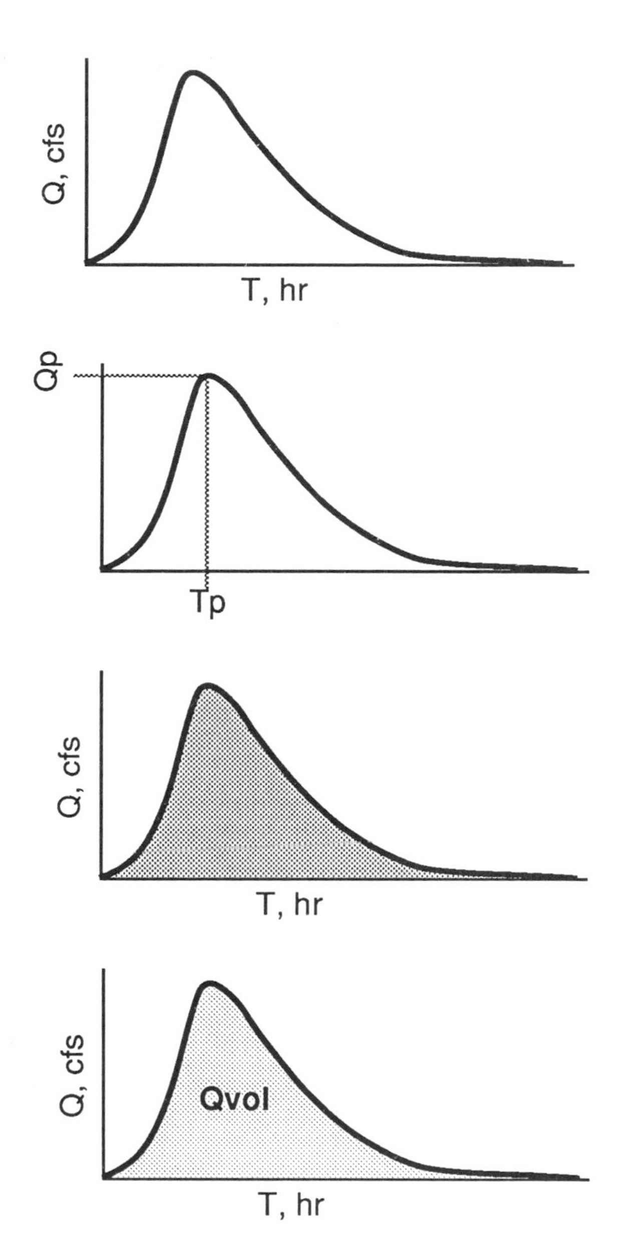

Let us look at a long-term stream hydrograph. Imagine that we stand by a perennial stream, and watch the rate of flow wax and wane for a period of weeks or months.

The general pattern of a long-term hydrograph is of long, low, more or less steady "base" flows, punctuated by relatively short, intense bursts of stormwater runoff. This makes sense in our personal experience because individual rain storms tend to last not more than a day or two each and to be followed by dry periods possibly weeks in length before the next storms come along.

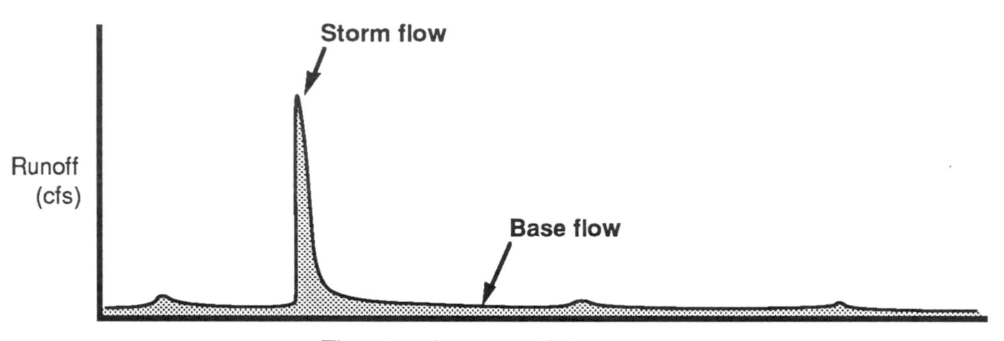

Storm flows are important to people because they are the occasions of flooding and stream erosion. Storm flows also go by the name of *peak flows* or *flood flows*. They need to be kept *low* in order to keep down flooding and erosion.

Base flows are connected to a wealth of additional resources and issues: surface water supplies, groundwater levels, assimilation of stream pollution, habitat of aquatic life, water recreation and aesthetics. Low flows are called *base flows* because on the hydrograph they seem to form a low plateau on which the spikes of storm flows are superimposed. (There is in fact a gradual decline of flow after each storm, giving a slight slope to the low-flow line on the hydrograph.) Base flows need to be kept *high* in order to maintain the resources that depend on them.

Both types of flow are issues. It is not possible to manage a stream in such a way as to make stream flow completely uniform all the time, because natural rainfall and evapotranspiration, which drive the stream system, vary from time to time. But we have to prevent development's aggravation of the magnitudes of the peaks and troughs.

To respond to these issues, there are two families of stormwater design. A *design storm* is an isolated, brief, rare and extreme peak-flow event. The *long-term water balance* encompasses low-level, continuous flows over a period of months or years, including, among others, evaporation, seepage and base-flow runoff. Both types of flows pass through every channel and every reservoir. The two approaches supplement each other.

Water quality is a relatively newly recognized but vital concern in stormwater. Numerous constituents can determine water quality. They may be in dissolved (chemical) or suspended (physical, particulate) form.

Harm from some constituents can result directly to humans when they reach water supplies or recreational lakes or are concentrated in fish, such as toxic chemicals and disease-related microorganisms. Others can damage wildlife and aquatic ecology. Coarse sediment can simply clog up storm sewers and streams, and prevent light from penetrating to aquatic plants. The degree of harm depends on the type and quantity of constituent.

Concentration is the ratio of constituent to water. A common unit is weight of constituent per volume of water, milligrams per liter (mg/l). Since in the metric system the volume and mass of water are linked, one mg/l at ordinary temperature is equal to one part per million (ppm, 0.1 percent by weight). See Appendix G for convertions between metric and American units.

The total amount of constituent is equal to the volume of water times the concentration:

Mass in mg = (Concentration in mg/l) (Volume in l)

The flow rate of a constituent is equal to the total amount of constituent divided by the period of time over which the flow occurs:

Flow rate in mg/day = Mass in mg / Time in days

Combining some of the above concepts, flow rate of a constituent is also equal to the concentration times the flow rate of the water that carries it:

Flow rate in mg/day = (Concentration in mg/l)(Water's flow in l/day)

The first flush during a storm is a particular water-quality concern. In regions where rainstorms are often separated by a few days or more, oil and other pollutants can accumulate on pavements between storms. When a rain starts, the first flush of water carries off most of the pollutants, and has a high concentration compared to runoff later in the storm. In these regions, constituent loads are associated not as much with quantity of runoff as with timing during each storm event.

Water carries a variety of constituents, the composition and quantity of which are influenced by climate, season, watershed mineralogy, constituents carried in by precipitation, and activities of man. There is no chemically pure water in nature. Here are some types of constituents that tend distinctively to characterize urban watersheds, and their sources. For discussions of their chemistry, biology, and ecological and health effects, see references listed in the *Bibiliography*.

Suspended solids:

Eroding soil;

Particles of glass, asphalt, stone, rubber and rust from cars and decomposing pavements.

Organic compounds:

Biological and chemical oxygen demand (BOD and COD), from nutrients;

Petroleum hydrocarbons from cars' oil and gasoline, concentrated at highways and big shopping-center parking lots;

Herbicides and pesticides from golf courses and other intensely maintained landscapes.

Nutrients:

Nitrogen and phosphorus from fertilizers concentrated around golf courses, certain residential areas, and other intensely maintained landscapes;

Wastes and leachates from dumpsters and service areas where trash is handled;

Automobile-related organic compounds, concentrated around highways and big shopping-center parking lots.

Trace metals:

From cars and trucks; lead can be high at highways and big shopping-center parking lots, but is becoming less important as leaded gas is phased out;

Decomposition of pavement and construction materials, including rusting iron, and zinc from galvanized roofs and culverts, nails, and painted surfaces;

Some fertilizers.

Chloride:

From deicing salts; high at highways and parking lots in winter in northern states.

Bacteria:

From animals, in warm season; may be concentrated around restaurants, dumpsters and service areas where trash is handled.

With that background we are ready to start considering stormwater management alternatives and doing the calculations to design for them.

Estimating hydrologic flows on a development site is the logical place to start. Any estimate of runoff is not a design; it is only an estimate of a natural process. But it is done for selected points on a site, under explicit assumptions of the conditions that may occur, so that we can deal with the resulting flows through design. Flows can be estimated for short-term peak storm events, or for long-term background flows. The types of hydrologic analysis you do should be aimed at the types of flow processes that you want to control through design.

Design alternatives are the remaining steps. These are qualitatively different management functions or scenarios. They are used to implement entirely different types of objectives for stormwater control and conservation. Every design alternative has some effect on both flow of water and flow of constituents. You have to design specifically for the types of effects you want to create. We will look at the alternatives in order of their historical development, from the most ancient to the most current and experimental.

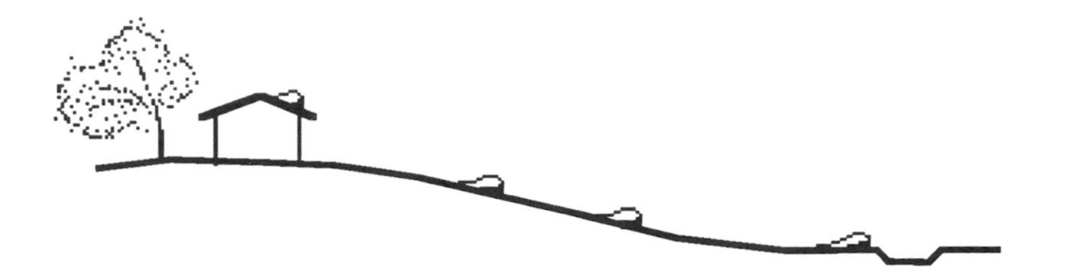

Conveyance is design for moving the water away. When conveyance is carried out without modification, it results in disposal of stormwater in surface streams or lakes without holding water back in any way.

The basic facilities for conveyance are pipes and channels. Even pipes that are buried under the ground are part of surface systems, because their impervious sides and bottoms prevent water from infiltrating the surrounding soil and allow it to move only laterally as all surface water does.

Conveyance is surely an ancient practice. If you visit the ancient Roman city of Pompeii you can see streets with gutters that drain continuously to discharge points.

In the modern industrial world Frederick Law Olmsted was a pioneer of stormwater conveyance in 1869, when he implemented it in the new community of Riverside, Illinois. Those were the days before automobiles, and the streets where people walked were covered with mud and horse manure. Olmsted said there was a nuisance problem, an aesthetic problem, a public health problem. He said the better alternative was to drop the stormwater and all the filth it carried off the streets, into a system of buried pipes. You can still go to Riverside, see the stormwater inlets he installed in the roads, and trace the pipes to their discharge in the Des Plaines River. Conveyance solved the problem of Olmsted's time.

More than a century later this type of management is still regularly taught in design and technical schools. Conveyance continues to be needed for draining land uses, discharging flood flows and allowing runoff from off-site watersheds to pass by our develoments without doing damage.

Some water constituents can accumulate, slowly, in the soil at the bottom of conveyance swales. But conveyance should not be counted as a major process for treating stormwater quality. By definition, the function of conveyance is to pass things through, not to accumulate them.

In today's world we know more about the effects of our developments on water quality, groundwater and rivers, and we are just as concerned about what happens to the filth in the Des Plaines River as the health of residents of our own developments. These concerns have caused us to consider alternative scenarios for stormwater management.

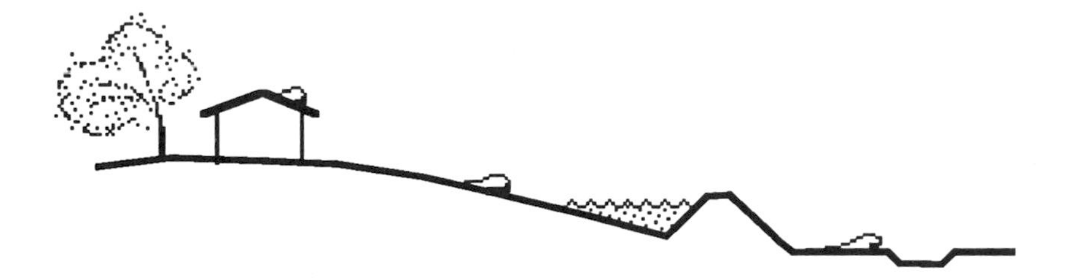

Detention is design for slowing water down as it moves away. The basic facility is a storage reservoir, either a permanent pond or a "dry" basin.

For flood control, short-term detention of storm flows has been widely practiced in metropolitan areas of the United States since about 1970. Storm detention responds to an increase in peak flood flow which was found in the 1960s to accompany urban development. Although its application to urban development is only a few decades old, it builds upon experience in regional flood control which goes back to the turn of the century. It tries to restrain flooding downstream from developments by reducing the rate at which surface water flows out. Although it reduces runoff's peak flow rate (*Qp*), the total volume of flow (*Qvol*) is still allowed to run downstream, only at a slower rate, stretched out over time.

Storm detention is a relative, quantitative modification of conveyance because it is still qualitatively a disposing of water at the surface. It is capable of suppressing flood peaks, when it is properly applied on a watershed-wide basis. However, storm detention's inability to adequately address water quality, ground water or water supplies has caused still other alternatives to be considered.

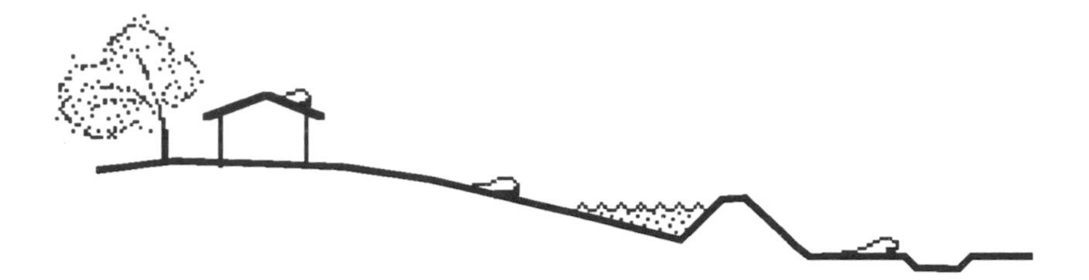

For water quality control, extended detention is becoming popular. When water sits still in a permanent pond or wetland, suspended particles can settle out, and chemicals can be adsorbed (adhered onto particle surfaces) in bottom sediment, taken up by plants, and transformed into gases that are released to the atmosphere. Extended detention requires a permanent pond larger than the expected volume of runoff; a "dry" pond will not do. A pond for treatment may be much larger than one needed for flood control alone, to give constituents adequate residence time for the treatment processes to be completed.

A pond or wetland is a land-based "treatment plant," although it is not as mechanical-looking as a sewage treatment plant. Stormwater's rates of flow are much more variable than those of sewage, so it is uneconomical to invest in a mechanical treatment plant big enough to handle peak short-term flows and then leave it unused for weeks at a time while lower flows trickle through. Land-based systems take advantage of the free, natural filtering and transforming capacities of light, air, soils and organisms. Although they require available land, they do not, relatively speaking, require a lot of expensive artificial materials or maintenance.

Like storm detention, extended detention is a relative, quantitative modification of conveyance. Essentially all runoff is eventually disposed into surface streams. Extended detention is capable of improving surface water quality, when it is specifically designed for this purpose. However, its inability to adequately address volume of runoff, ground water or water conservation leads us to consider still other alternatives.

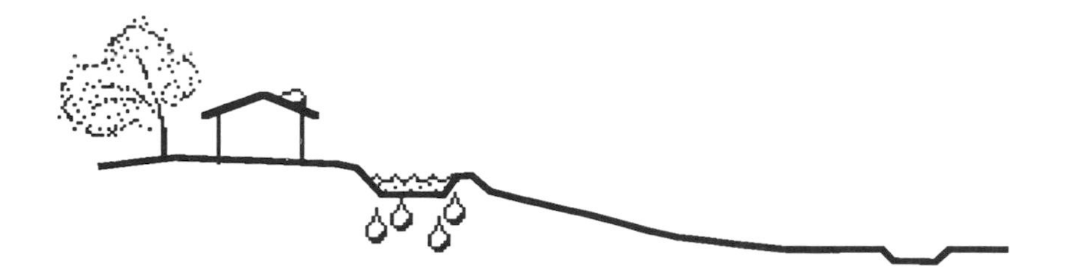

Infiltration is design for soaking water into the ground. The water no longer moves laterally across the surface. Infiltration is qualitatively different from conveyance and detention because it makes water go to a different place, where it undergoes new types of processes. Infiltration addresses flooding (as does detention) as well as water quality, groundwater and water supplies. Although it is still a form of stormwater disposal (because it releases water to the environment) it puts the water in a part of the environment where it can be filtered, stored and available for further use. Unlike other alternatives, it restores natural hydrologic processes and manages long-term groundwater levels and stream base flows.

The basic facility is an earthen reservoir with no regular surface outlet. The basin can be "dry" if on well-drained soil, "wet" if in contact with permanent ground water, or "ephemeral" if the water level is allowed to fluctuate widely.

Experience with infiltration has been gained for half a century on Long Island, where it goes by the name of *recharge* since it feeds underlying aquifers. In other parts of the country, notably Florida and certain parts of Arizona, California, New England and the mid-Atlantic states, its practice has fluctuated as various programs have been tried. Since infiltration works with relatively new and unfamiliar types of processes, it is currently the subject of research and evaluation.

Infiltration can improve water quality. Soils can filter out suspended particles and adsorb chemicals. The ecological processes in soils can transform constituents to gases that are released to the atmosphere. Almost all of the filtering and transformation processes take place in the top few inches of soil, without carrying constituents into underlying ground water. Like an extended-detention pond, an infiltration basin is a kind of treatment plant.

It is an exciting time for this new alternative, which may supplement detention as a major stormwater management scenario in many areas of the country.

Water harvesting is design for capturing and using stormwater runoff on-site. One could say that infiltration allows potential water reuse by storing water in accessible parts of the environment. In contrast, water harvesting picks up and uses runoff in the development site where it is generated.

For permanent ponds and wetlands, water harvesting can supply runoff to maintain water levels. During the dry months, enough can be supplied to balance a pond's water level. In wetter periods, surplus runoff water would pass through a pond's outlet.

For dry basins, water harvesting can be turned around to assure that runoff supply is dispersed enough over a basin's floor *not* to accumulate as standing water, even during the wetter months. The desired multiple uses and landscape effects of the basin can be assured.

For landscape irrigation, water harvesting can supplement artificial irrigation systems. A simple way is to let runoff soak into the soil in infiltration basins where water-loving plants are rooted. A more elaborate way is to store runoff for days or weeks at a time in reservoirs, then pump it into the normal irrigation lines.

Water harvesting is an ancient practice. It can be seen on the rooftops and urbans plazas of cities of classical and medieval civilizations.

In the modern industrial world this scenario has only recently been rediscovered and adapted to our new way of life. Ponds have been built in urban developments for a long time, but now they can be analyzed before construction to find out whether an adequate runoff supply is in fact feasible. In Arizona and Colorado water harvesting for irrigation is passing out of the experimental stage to be applied widely to urban landscapes. As our demands upon water become more critical and as we become still more sensitive to the environmental impacts of land development, water harvesting may become an important part of our future.

Chapter 2

Storm Runoff Estimation

Estimation of storm runoff defines the volumes of water and rates of flow in individual peak-flow events. It is a modeling of a natural process.

Before design to control peak-flow events can be done, the natural process that we are dealing with must be estimated. Many of the design applications covered in this book are founded upon estimates of individual runoff events. We often have to estimate flow volume and peak flow rate both before and after development in order to deal with the impacts of land use change.

At a specific site, a runoff gauging station would be a direct way to observe flows from the site in its existing predevelopment condition. But few development sites have gauging stations. Even if a gauging station were present, it would not predict the flows that would occur after a proposed development is constructed.

Some sort of estimate is necessary, based on general knowledge of runoff processes and relevant data about the site.

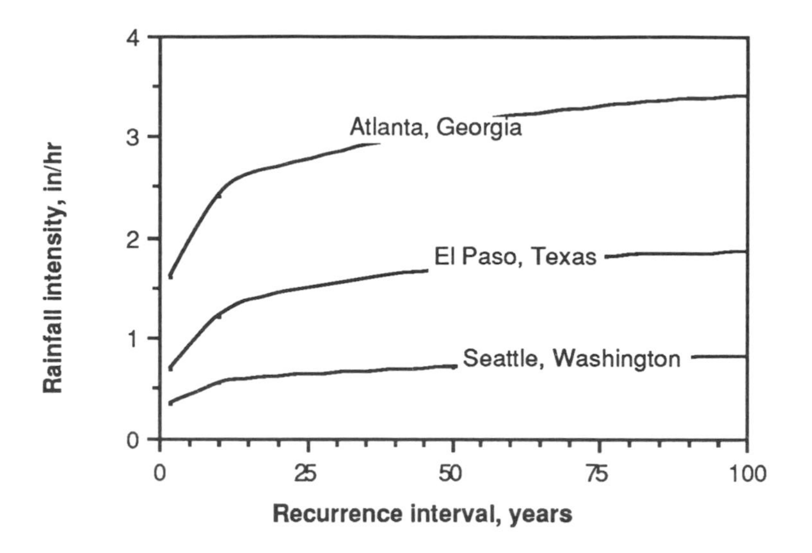

Rainfall-runoff models estimate runoff based on the *drainage area* that outlets at a point and the *rainfall* upon it. They are convenient methods because there are rain gauges at many local weather stations around the country, and it is presumed that rainfall records from a weather station are valid for all sites in the surrounding region.

A design storm is a particular combination of rainfall conditions for which you estimate runoff.

The magnitude of a design storm might be expressed as a total quantity of precipitation, such as inches of rainfall, or as a short-term intensity, such as inches per hour. A channel or basin designed for a storm of a certain size is expected to overflow only when a storm larger than the design storm occurs.

Recurrence interval is a way to express the probability that a given type of storm might occur at your site. Recurrence interval is the average time between storms of a given magnitude in a local rainfall record. A *10-year storm* is large enough that it has recurred, on the average, in one of every 10 years in the record; it has a 10 percent chance of occurring in any one year. A *100-year storm* is so big that it has occurred only once every 100 years; it has a 1 percent chance of occurring in any one year. The probability of occurrence in any one year is the reciprocal of the recurrence interval.

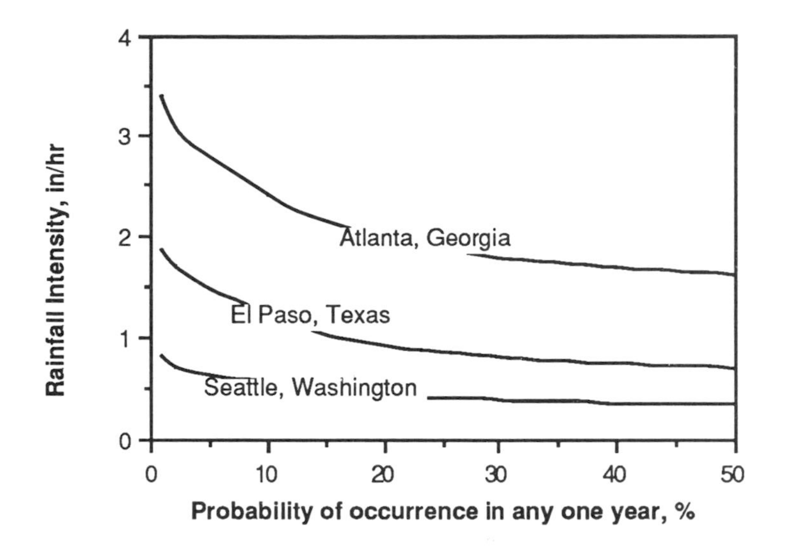

There is always a risk that a design storm that you have selected could be exceeded and stormwater facilities you have designed could be overloaded. The question is what balance of risk and cost is appropriate to each specific project. Selection of a small, frequent design storm would allow small, inexpensive facilities, but would cause rather frequent overflows. Selection of a larger, less frequent design storm would result in facilities that contain the runoff they receive almost all the time, but that are larger and more expensive. Local ordinances may govern the recurrence interval to be used. Somewhere around a 2 to 50 year recurrence interval is common for many types of on-site stormwater facilities. In extremely sensitive situations, where people's homes would be seriously damaged by flooding, the most appropriate storm is the *maximum probable* storm, which can be much larger than the 100-year storm.

Two drainage systems exist side-by-side on every site, the *primary* system and the *secondary*. The primary system operates during all storms up to and including the design storm. The secondary system begins to operate whenever the primary system is clogged or a larger storm occurs.

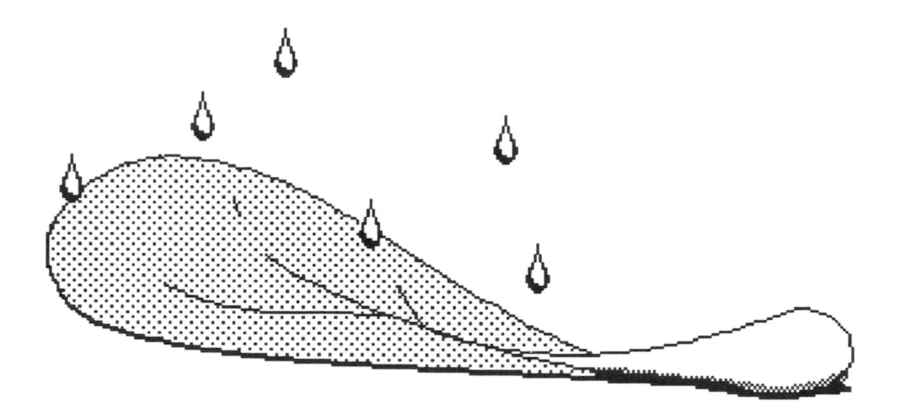

The drainage area is the land area that drains to the point where you estimate runoff. Any rainfall-runoff model requires you to specify certain characteristics of the drainage area. See Appendix B if you are not familiar with procedures for identifying drainage areas or analyzing their basic characteristics.

Runoff's travel time is one watershed characteristic that can strongly influence the rate of discharge. The faster a given volume of runoff drains off, the greater the peak rate of flow at the outlet.

Travel time can be expressed by water's velocity and the distance it has to travel:

$$Tt = L / V$$

where,

Tt = travel time in minutes;
L = length of travel in feet: can be scaled from a site map; and
V = velocity in feet per minute: from the table on the next page, or using the procedures discussed in Appendix E.

If the flow path traverses more than one kind of surface or slope with different velocities, the individual travel times across them should be added up.

Flow Velocities

For slopes and surfaces not listed here,
use the value of the most similar listed slope and surface,
or interpolate between pairs of values shown.

	Flow velocity, feet per minute (fpm)	
Land surface	Slope about 100 feet in length	Slope up to 500 feet in length
Paved slope	20	50
Bare area	15	30
Turf	10	15
Rangeland hillslope	1.2-3.0	1.2-3.0
Small natural channel: Derived from Manning's equation	360 max.	360 max.
Artificial channel: Derived from Manning's equation	480 max.	480 max.

Derived from page 302 of Dunne and Leopold, 1978, *Water in Environmental Planning*.

Time of concentration is a special case of travel time. It is the *maximum* time runoff from any point in a drainage area could take to flow to the outlet. Among a number of alternative paths and travel times that runoff could take from distant parts of a watershed, the time of concentration is defined by the longest possible time, whether or not it involves the longest distance measured horizontally. The line along which that runoff travels is the drainage area's *hydraulic length.* The most common storm runoff models use watershed time of concentration as their indication of how fast water could drain off a site and contribute to peak flow of runoff.

For simple manual estimation of time of concentration, you can apply the time-of-travel equation $T = L/V$ directly. Scale hydraulic length Hl from a site map. Obtain velocity V from the table on page 23. Time of concentration Tc is then found from substituting into the travel-time equation, giving

$$Tc = Hl / V.$$

The Soil Conservation Service method of calculating runoff, as set forth in the 1986 edition of SCS's *Technical Release 55*, suggests breaking down the hydraulic length into two or three segments for estimating travel times. A separate approach is used to find the travel time in each segment, then the times are added up to give total time of concentration. It is a laborious method. Although we do not take your time to describe it fully in this chapter, you should be aware of it if you will need to work under a strict interpretation of *TR55*. The method is described in our Appendix E.

The computer disk that accompanies this book uses a slightly different way of finding watershed time of concentration, but is still valid under the SCS method. When using the disk, you specify the total hydraulic length in feet, and the elevation along the hydraulic length at one point 10 percent of the length from the bottom, and at another point 85 percent of the way from the bottom. You also specify percent of the hydraulic length that is channelized, and other information about the watershed's soil and land use. The computer calculates slope and velocity, and thus time of concentration, from the information you give it. When you have a downstream network of drainage pipes or channels, you specify velocity in each segment of the network. The computer accumulates time of travel and time of concentration through the drainage system, and uses them in calculating downstream hydrographs. If you prefer to use the *TR-55* time-of-concentration estimate, or any other estimate, the computer model allows you to enter your estimated Tc at an appropriate point during the running of the model.

> ***A note about exercises***
> This book includes an exercise for each type of procedure that is discussed. Exercises later in the book use results of earlier ones. We have found that solving each exercise for two sites that contrast in some way such as soil or rainfall stimulates discussion and understanding of hydrologic and land use factors. You can use sites that you are working on in practice, typical sites in your region with which you are familiar, or hypothetical sites described in Appendix A.

Time of Concentration Exercise

		Site 1	*Site 2*
		1. Before Development	
Hydraulic length *Hl* (from site map)	=	______ ft.	______ ft.
Material along hydraulic length (from site map)	=	______	______
Velocity of travel *V* (page 23)	=	______ fpm	______ fpm
Time of concentration = *Hl* / *V*	=	______ min.	______ min.
		2. After Development	
Hydraulic length *Hl* (from site map)	=	______ ft.	______ ft.
Material along hydraulic length (from site map)	=	______	______
Velocity of travel *V* (page 23)	=	______ fpm	______ fpm
Time of concentration = *Hl* / *V*	=	______ min.	______ min.

Runoff Estimation:
The "Rational" Method

The "rational" method is about 100 years old. This rainfall-runoff model seems to have come to the United States during the early development of New York, where newly installed storm sewers were frequently overloaded and a "rational" method for sizing conveyances was called for. The method's simple equation did provide such a method for the first time. It is still widely used for on-site stormwater facilities because it is so simple to use. Since the "rational" method was developed to estimate peak flows from small drainage areas, its application today is usually restricted to drainage areas of less than 200 acres.

The "rational" method associates peak rate of runoff with three easily identifiable characteristics of a drainage area and the rainfall upon it. The variables are:

A = drainage area (acres):
can be scaled from a site map.

C = drainage area cover factor (no units),
based on a combination of soil, land use, and slope: see the table on page 27.

I = rainfall intensity (in/hr) at a selected recurrence interval and a storm duration equal to the time of concentration: see the graphs on pages 29 and 30, and in Appendix A.

A simple formula connects these variables: with Qp being peak rate of runoff in cfs,

$$Qp = C\,I\,A$$

Keep the units straight: multiplying CIA does not automatically give cfs. In the early days there was an additional factor for correcting the units, equal to 1.008 cfs/ac.in.hr. Since the factor made no appreciable difference to the result eventually everybody just forgot about writing it into the formula. Now we are left with a very short, simple formula, but one where we have to make sure we use the correct units for each variable.

The rational method was intended, when originally developed, only to identify the peak rate of flow that those pipes and culverts had to carry. It estimates intensity of runoff in cfs directly from intensity of rainfall in in/hr. It was not intended to estimate total volume of flow during the storm, although some workers have added supplemental assumptions to "stretch" the method to give volume. For applications that involve volume of flow the SCS method is preferable, since it was originally developed to include it.

Cover Factors

For flat slopes or permeable soils use the lower values.
For steep slopes or impermeable soils use the higher values.

Type of surface or land use	Cover factor *C* in rational formula
Forest	0.1 - 0.3
Turf or meadow	0.1 - 0.4
Cultivated field	0.2 - 0.4
Steep grassed area (2:1)	0.5 - 0.7
Bare earth	0.2 - 0.9
Pavement, concrete or asphalt	0.8 - 0.9
Flat residential, about 30% impervious	0.40
Flat residential, about 60% impervious	0.55
Sloping residential, about 50% impervious	0.65
Sloping, built-up, about 70% impervious	0.80
Flat commercial, about 90% impervious	0.80

From Table 1 in U.S. Federal Highway Administration, 1973, *Design of Roadside Drainage Channels.*

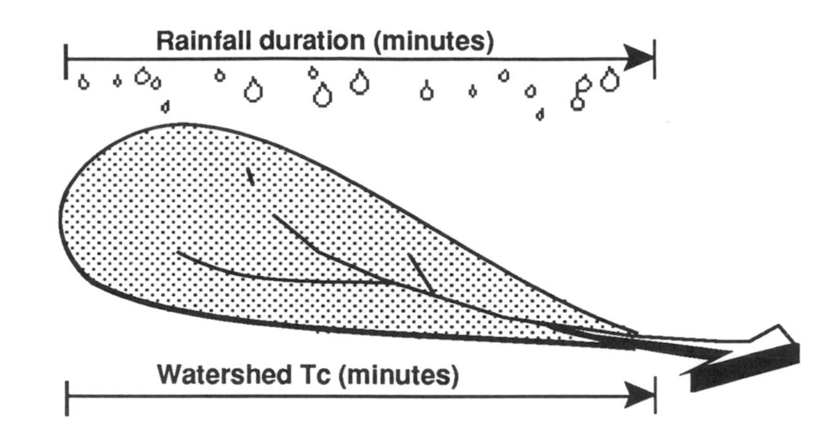

Storm duration is a concern in the "rational" method because the method presumes that, at a given recurrence interval, a storm *duration* which *matches* a drainage area's time of concentration can produce the greatest peak rate of runoff for a given recurrence interval. The idea is that a *shorter* duration would indicate a more intense storm, but the rain would stop before runoff from all of the watershed arrived at the outlet to contribute to peak discharge. A storm with *longer* duration would continue adding rainfall as long as the runoff keeps traveling to the outlet, but would be on the whole a less intense storm. A duration *exactly equal* to the time of concentration indicates the most intense storm that will keep on contributing rainfall as long as the watershed takes to concentrate runoff upon the outlet.

The following graphs are examples showing rainfall intensity as a function of storm duration and recurrence interval at selected locations. Graphs for other selected weather stations are in the appendix on *Exercise Sites*. Similar graphs are available for other locations in the U.S. Weather Bureau's Technical Paper No. 24, *Rainfall Intensity-Duration-Frequency Curves,* 1955, or local sources such as the article by Aron and others for Pennsylvania, cited in the *Bibliography*. For sites without specific rainfall charts, interpolate rainfall geographically from charts of nearby cities.

A number of different versions of the "rational" formula are being applied in practice. These variation have grown out of the formula's open-ended, exploratory beginnings and evolution over a century of widespread use. You can find many different personal judgments about which version is best. In calculating time of concentration some versions use only overland flow time while others use combinations of flow conditions similar to the SCS method. Several different tables of cover factors are used, some listing high values in order to generate conservative (high) peak discharge estimates. Different practitioners can get very different results for the same watershed, depending on their assumptions. Such variability in application can be considered a weakness of the "rational" method. Further details about the "rational" formula and all other procedures discussed in this book can be found in references listed in the *Bibliography*.

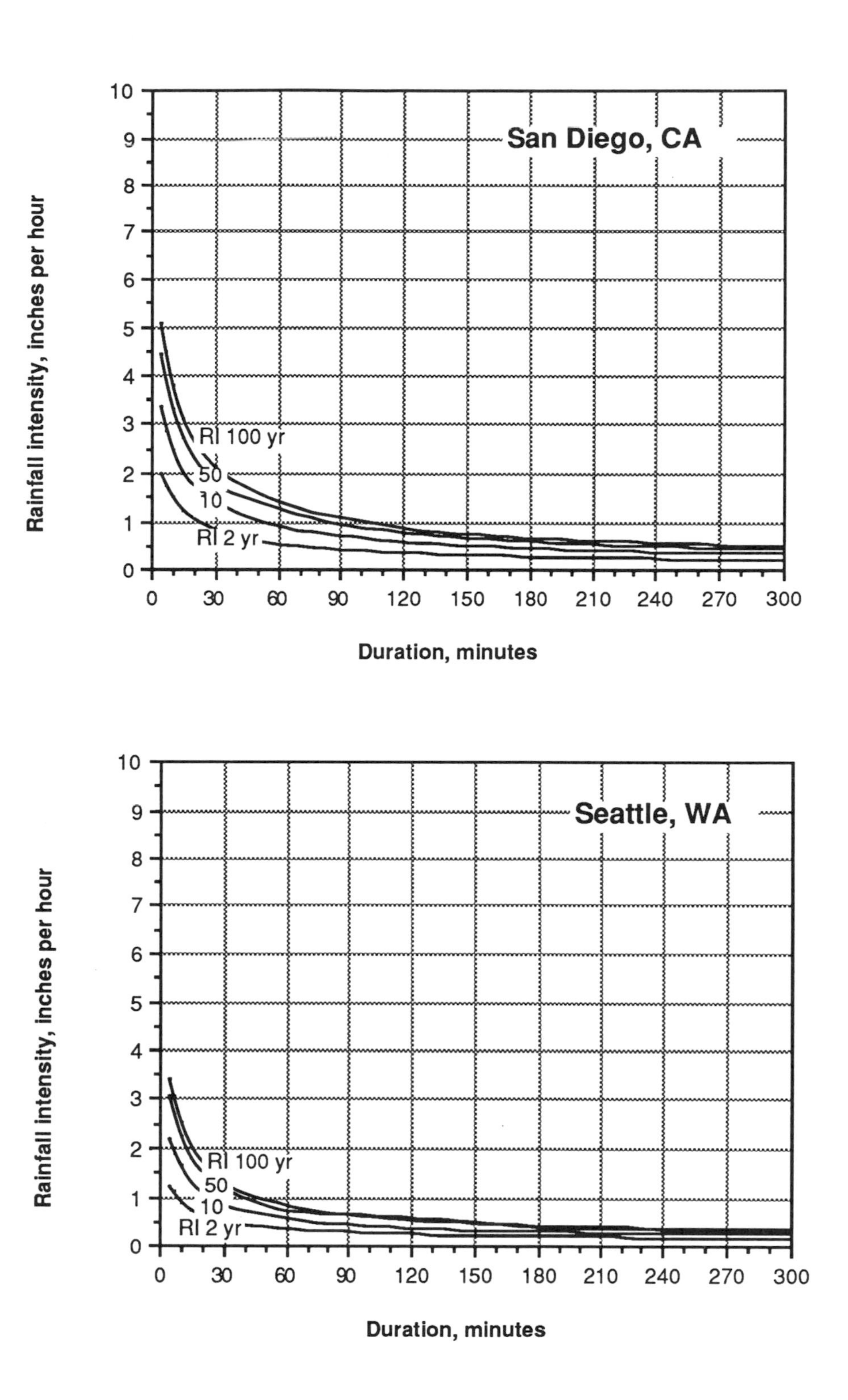

Charts on this page were derived from U.S. Weather Bureau, 1955, *Rainfall Intensity-Duration-Frequency Curves,* Technical Paper No. 25

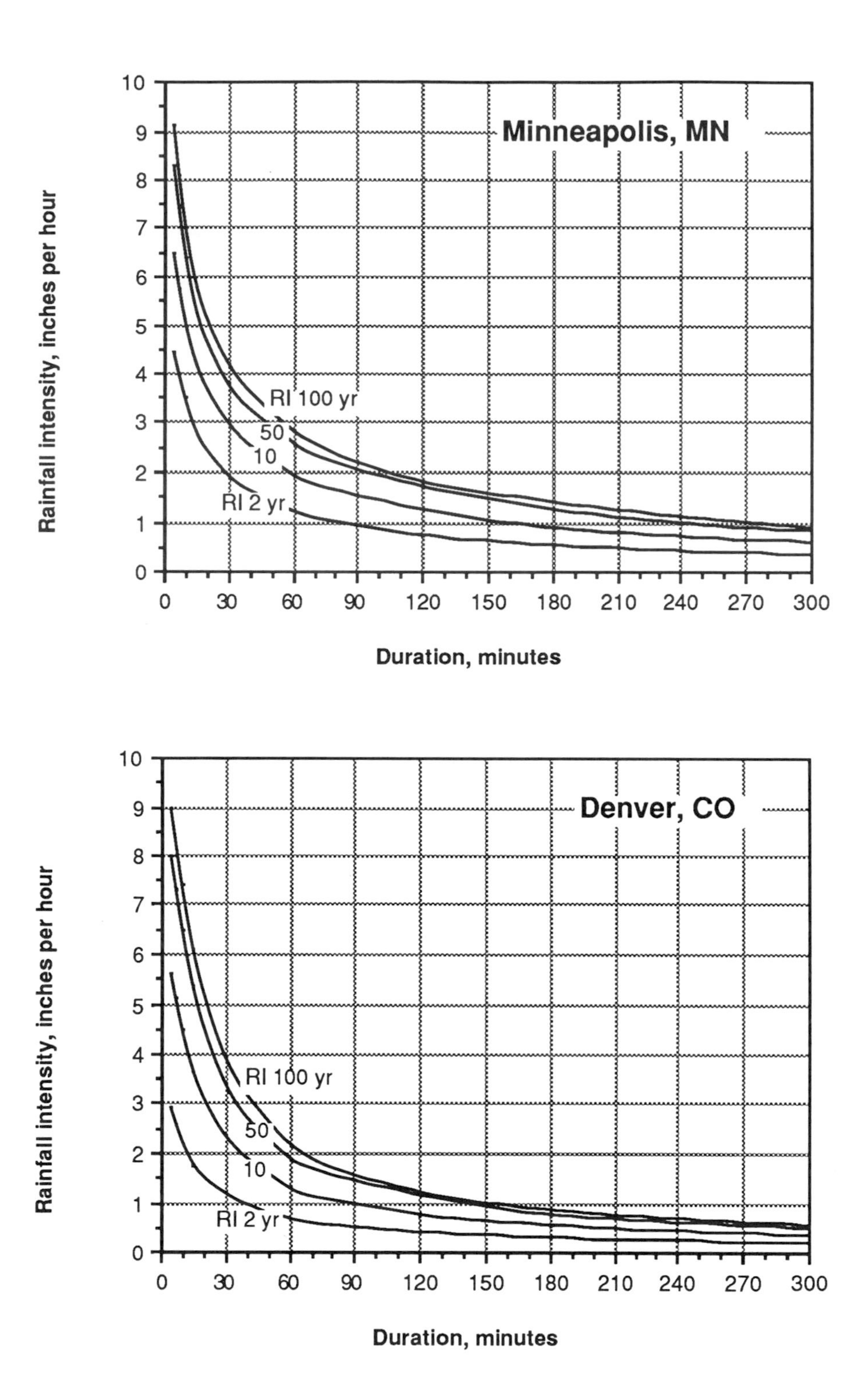

Charts on this page were derived from U.S. Weather Bureau, 1955,
Rainfall Intensity-Duration-Frequency Curves, Technical Paper No. 25

Summary of Process
"Rational" Formula

1. Obtain time of concentration *Tc* in minutes from page 25.

2. Obtain rainfall intensity *I* in in/hr, at a selected recurrence interval and duration equal to *Tc*, from one of the graphs on pages 29-30 or in Appendix A, U.S. Weather Bureau Technical Paper No. 25, or local data.

3. Obtain cover factor *C* from the table on page 27.

4. Estimate watershed area *A* in acres, from a site map.

5. Compute peak rate of runoff *Qp* in cfs from the equation,

$$Qp = C\,I\,A.$$

Runoff Estimation Exercise

Rational Method, After Development

		Site 1	*Site 2*
Time of concentration (from page 25)	=	______ min	______ min
Rainfall intensity *I* (from graph)	=	______ in/hr	______ in/hr
Cover factor *C* (from page 27)	=	______	______
Drainage area *A* (from site map)	=	______ ac	______ ac
Peak rate of flow *Qp* = *C I A*	=	______ cfs	______ cfs

Runoff Estimation:
The SCS Method

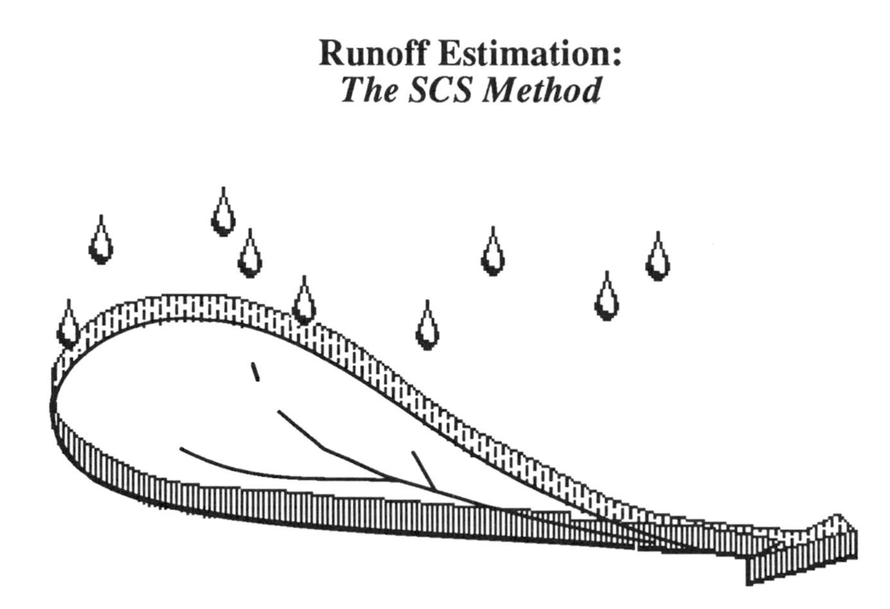

The SCS method was developed and documented entirely by the U.S. Soil Conservation Service (SCS), beginning in the 1950s. The SCS has published many handbooks explaining the method, all with slightly different charts and sequences of steps. Each one is a special-purpose interpretation of the original SCS documents, the most central of which are the *Engineering Handbook, Chapter 4, Hydrology*, and *Technical Release 20*, which is a computer implementation of the hydrologic formulas in the *Engineering Handbook*.The most prominent and useful of SCS's interpretive publications is *Technical Release 55* (*TR-55*), *Urban Hydrology for Small Watersheds*, which is the source for most of our discussion here.

The SCS method is growing in prominence, gradually replacing the "rational" method. It is not necessarily more accurate than the "rational" formula, where it has not been correlated to local conditions by additional procedures. But it is more directly useful for applications requiring an estimate of runoff volume, and its thorough documentation usually allows it to be used without a lot of disagreement among designers, planning commissioners, etc., about how it is supposed to be applied. It can be applied to drainage areas much larger than those to which the "rational" method is limited.

Its disadvantage has been that it is more complicated to learn and to apply than the "rational" method. Manual solutions of the original equations would be extremely time consuming; it should be applied only with interpretive charts or on a computer. With microcomputers becoming more common, the use of computer programs that reiterate the SCS method eliminates the argument about difficulty of applying all the original SCS equations. In this book we provide both charts and a computer disk incorporating the method. Appendix H is a manual for operation of the computer model.

Runoff volume in acre feet is established by the depth of storm runoff. Unlike the "rational" method, the SCS method first finds depth of runoff, based on the depth of storm rainfall.

The *rate* of runoff in cubic feet per second is then determined by how fast a given volume of runoff drains off a watershed (the time of concentration), and the distribution of rainfall intensity during the storm.

The equation for runoff depth is considered the method's basic equation (Equation 2-1 of *TR-55*):

$$Dr = \frac{(P - Ia)^2}{(P - Ia + S)}$$

where,

Dr = depth of runoff (in.);
P = depth of 24-hour rainfall (in.);
Ia = initial abstraction, the losses of rainfall before runoff begins (in.); 0.2 is a more or less median value that has been found in the field; and
S = potential maximum retention after runoff begins (in.); it is defined by Curve Number CN, which is a function of the drainage area's soil and land use:

$$S = (1000/CN) - 10$$

To find depth, and thus volume, of runoff, whether using the basic equation or interpretive charts, you first need specific information about rainfall and the watershed.

The SCS method uses a 24-hour rainfall, with the intensity of rain within that period assumed to be in a certain distribution. Twenty-four hour rainfall amounts at three recurrence intervals are shown in the following maps.

Small-scale maps like those in this book should be used with caution, particularly in mountainous areas where precipitation can vary greatly within a short distance. More detailed maps are available for the following regions:

• **For states west** of the 105th meridian: National Oceanic and Atmospheric Admistration (NOAA) Atlas 2, *Precipitation-Frequency Atlas of the United States*, by John F. Miller and others, 1973. The NOAA Atlas comes in separate volumes for the individual states.

• **For Alaska**: U.S. Weather Bureau Technical Paper No. 47, *Probable Maximum Precipitation and Rainfall-Frequency Data for Alaska*, by John F. Miller, 1963.

• **For Hawaii**: U.S. Weather Bureau Technical Paper No. 43, *Rainfall-Frequency Atlas of the Hawaiian Islands*, 1962.

• **For Puerto Rico and Virgin Islands**: U.S. Weather Bureau Technical Paper No. 42, *Generalized Estimates of Probable Maximum Precipitation and Rainfall-Frequency Data for Puerto Rico and Virgin Islands*, 1961.

2 year, 24 hour Rainfall, in Inches

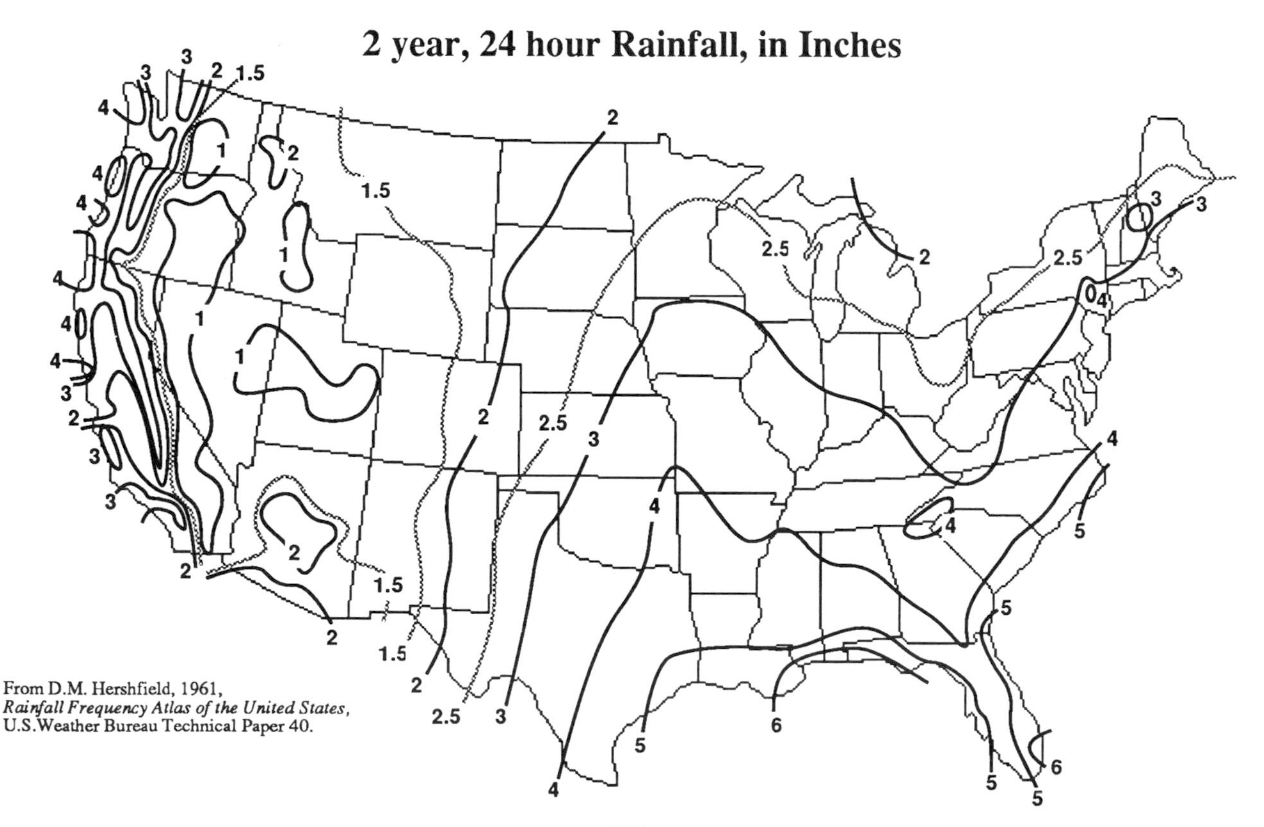

From D.M. Hershfield, 1961, *Rainfall Frequency Atlas of the United States*, U.S.Weather Bureau Technical Paper 40.

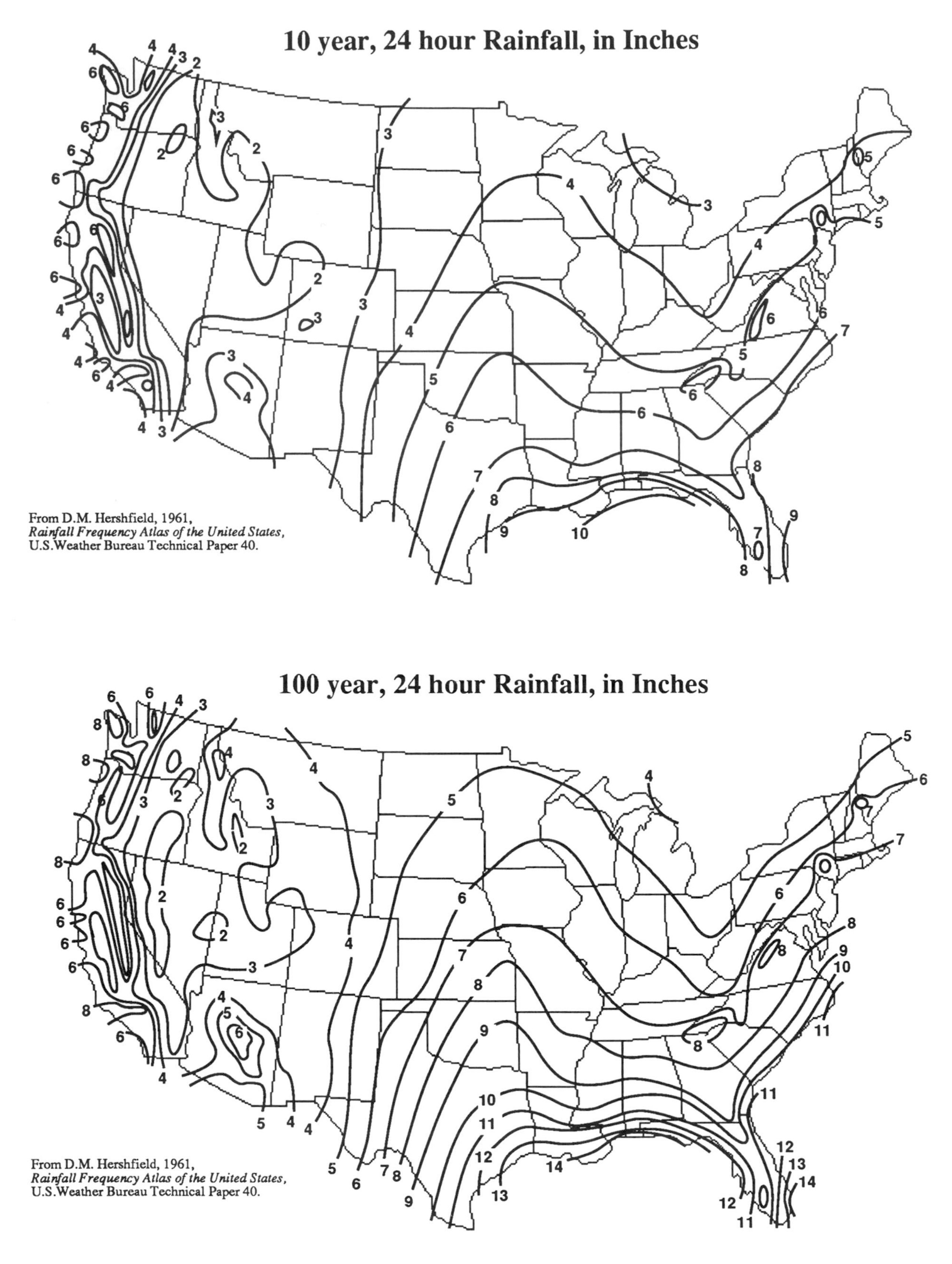
10 year, 24 hour Rainfall, in Inches
From D.M. Hershfield, 1961,
Rainfall Frequency Atlas of the United States,
U.S.Weather Bureau Technical Paper 40.
100 year, 24 hour Rainfall, in Inches
From D.M. Hershfield, 1961,
Rainfall Frequency Atlas of the United States,
U.S.Weather Bureau Technical Paper 40.

Hydrologic Soil Group (*HSG*) is SCS's way of summarizing soil's hydrologic effects. This classification, with land use, is one of the determinants of SCS's Curve Number. SCS has categorized every soil in the country into four groups, lettered A to D. Group A is the least likely to create runoff; group D is the most likely.

The four groups are defined by SCS soil scientists as follows:

• **Group A** soils have low runoff potential and high infiltration rates even when thoroughly wetted. They consist chiefly of deep, well to excessively drained sands or gravels and have a high rate of water transmission (greater than 0.30 in/hr). This group also includes sand, loamy sand and sandy loam that have experienced urbanization but not been significantly compacted.

• **Group B** soils have moderate infiltration rates when thoroughly wetted and consist chiefly of moderately deep to deep, moderately well to well drained soils with moderately fine to moderately coarse textures. These soils have a moderate rate of water transmission (0.15 to 0.30 in/hr). This group also includes silt loam and loam that have experienced urbanization but not been significantly compacted.

• **Group C** soils have low infiltration rates when thoroughly wetted and consist chiefly of soils with a layer that impedes downward movement of water and soils with moderately fine to fine texture. These soils have a low rate of water transmission (0.05 to 0.15 in/hr). This group also includes sandy clay loam that has experienced urbanization but not been significantly compacted.

• **Group D** soils have high runoff potential. They have very low infiltration rates when thoroughly wetted and consist chiefly of clay soils with a high swelling potential, soils with a permanent high water table, soils with a claypan or clay layer at or near the surface, and shallow soils over nearly impervious material. These soils have a very low rate of water transmission (0 to 0.05 in/hr). This group also includes clay loam, silty clay loam, sandy clay, silty clay and clay that have experienced urbanization but not been significantly compacted.

• **Compound classification A/D** indicates that the natural soil is in group D because of a high water table which impedes infiltration and transmission, but following artificial drainage using such methods as perforated pipe underdrains, the soil's classification is changed to A.

For a specific site, *HSG* designations can be obtained by referring to a local SCS soil survey where one is available. If the survey does not specify *HSG*s, you can look up the soil names in the complete national listing given in SCS's *Technical Release 55*. If there is no SCS survey at all, you can make an on-site investigation of soil characteristics, and compare them with the above definitions.

Information about HSG on this page is from page A-1 of U.S. Soil Conservation Service, 1986, *Urban Hydrology for Small Watersheds*, Technical Release 55.

Curve Number (*CN*) summarizes the effects of soil and land use, analogous to the cover factor in the "rational" formula. Curve number has no units. It varies from 0 to 100 as the ability of soil and cover conditions to generate runoff increase.

The strange name for this factor comes from curves on the graphs that SCS used in developing its method. Every combination of hydrologic soil group and land use type has a separate curve on the graphs. Each curve has, literally, a "curve number."

The tables on the next two pages summarize SCS's *CN* designations for specific combinations of *HSG* and land use.

Composite ***CN*****s** may need to be calculated for areas with combinations of surface cover, such as the combinations of grass and woods in parks and cemeteries. To do so find the *CN*s for the individual types of surface cover, then average them, weighting them by the proportion of the total area they cover. If you need a reminder of how to do a weighted average see the Appendix on *Drainage Area Analysis*.

Urban areas involve special kinds of composite *CN*s. Urban areas consist of combinations of pervious and impervious surfaces, with some of the impervious area directly connected to the drainage system and some not. Impervious areas are "directly connected" to the drainage system if runoff from them flows directly via concentrated flow into the site's drainage system. They are "unconnected" if runoff from them spreads as sheet flow over pervious areas before entering the drainage system. Obtain the *CN* of a pervious surface cover from the tables. Then find the composite urban *CN* using the chart on page 41 that is most appropriate for your site's degree of connectedness to the drainage system.

If an urban area has a combination of connected and unconnected impervious areas, find the proportion of impervious area that is in each condition, apply each proportion to the appropriate chart, and interpolate linearly between the results of using the two charts separately.

Curve Numbers in Urban and Agricultural Areas

Surface cover	*Hydrologic Soil Group*			
	A	*B*	*C*	*D*
Meadow:				
Continuous grass, not grazed, generally mowed for hay	30	58	71	78
Lawns, turf:				
>75% grass cover	39	61	74	80
50 to 75% grass cover	49	69	79	84
<50% grass cover	68	79	86	89
Pasture, grassland or range in humid regions:				
>75% ground cover and lightly or occasionally grazed	39	61	74	80
50 to 75% grass cover	49	69	79	84
<50% ground cover or heavily grazed with no mulch	68	79	86	89
Small grains:				
Contoured & terraced, high density, ≥20% residue cover	59	70	78	81
Contoured & terraced, low density, <20% residue cover	61	72	79	82
Straight rows, high density, ≥20% residue cover	63	75	83	87
Straight rows, low density, <20% residue cover	65	76	84	88
Row crops:				
Contoured & terraced, high density, ≥20% residue cover	62	71	78	81
Contoured & terraced, low density, <20% residue cover	66	74	80	82
Straight rows, high density, ≥20% residue cover	67	78	85	89
Straight rows, low density, <20% residue cover	72	81	88	91
Farmsteads:				
Buildings, lanes, driveways, and surrounding lots	59	74	82	86
Desert-shrub landscaping:				
Natural desert landscaping, pervious areas only	63	77	85	88
Impervious weed barrier and mulch, with basin borders	96	96	96	96
Fallow:				
≥20% residue cover	74	83	88	90
<20% residue cover	76	85	90	93
Bare soil	77	86	91	94
Newly graded areas:				
Pervious areas only, no vegetation	77	86	91	94
Streets, roads, buildings, structures:				
Dirt roads, including right-of-way	72	82	87	89
Gravel roads, including right-of-way	76	85	89	91
Paved road with open ditches, including right-of-way	83	89	92	93
Impervious roofs and pavements	98	98	98	98

From Table 2-2 of U.S. Soil Conservation Service, 1986, *Urban Hydrology for Small Watersheds*, TR 55, second edition. Values in this table assume that antecedent runoff condition (the runoff potential before a storm event) is average, that vegetation is fully established, and that $Ia = 0.2S$.

Curve Numbers in Natural Areas

Surface cover	*Hydrologic Soil Group*			
	A	*B*	*C*	*D*
Woods:				
No grazing; litter and brush cover the soil	30	55	70	77
Grazed but not burned, some forest litter covers soil	36	60	73	79
Heavy grazing or regular burning destroy litter, brush	45	66	77	83
Desert shrub (saltbush, mesquite, creosotebush, etc.):				
>70% ground cover	49	68	79	84
30 to 70% ground cover	55	72	81	86
<30% ground cover (litter, grass, and brush overstory)	63	77	85	88
Brush, grass and weeds in arid and semiarid regions:				
>70% ground cover	?	62	74	85
30 to 70% ground cover	?	71	81	89
<30% ground cover (litter, grass, and brush overstory)	?	80	87	93
Pinyon & juniper with grass understory:				
>70% ground cover	?	41	61	71
30 to 70% ground cover	?	58	73	80
<30% ground cover (litter, grass, and brush overstory)	?	75	85	89
Brush, grass and weeds in humid regions:				
>75% ground cover	30	48	65	73
50 to 75% ground cover	35	56	70	77
<50% ground cover	48	67	77	83
Oak-aspen mountain brush (oak, aspen, bitter brush, etc.):				
>70% ground cover	?	30	41	48
30 to 70% ground cover	?	48	57	63
<30% ground cover (litter, grass, and brush overstory)	?	66	74	79
Sagebrush with grass understory:				
>70% ground cover	?	35	47	55
30 to 70% ground cover	?	51	63	70
<30% ground cover (litter, grass, and brush overstory)	?	67	80	85

From Table 2-2 of U.S. Soil Conservation Service, 1986, *Urban Hydrology for Small Watersheds*, TR 55, second edition. Values in this table assume that antecedent runoff condition (the runoff potential before a storm event) is average, that vegetation is fully established, and that $Ia = 0.2S$.

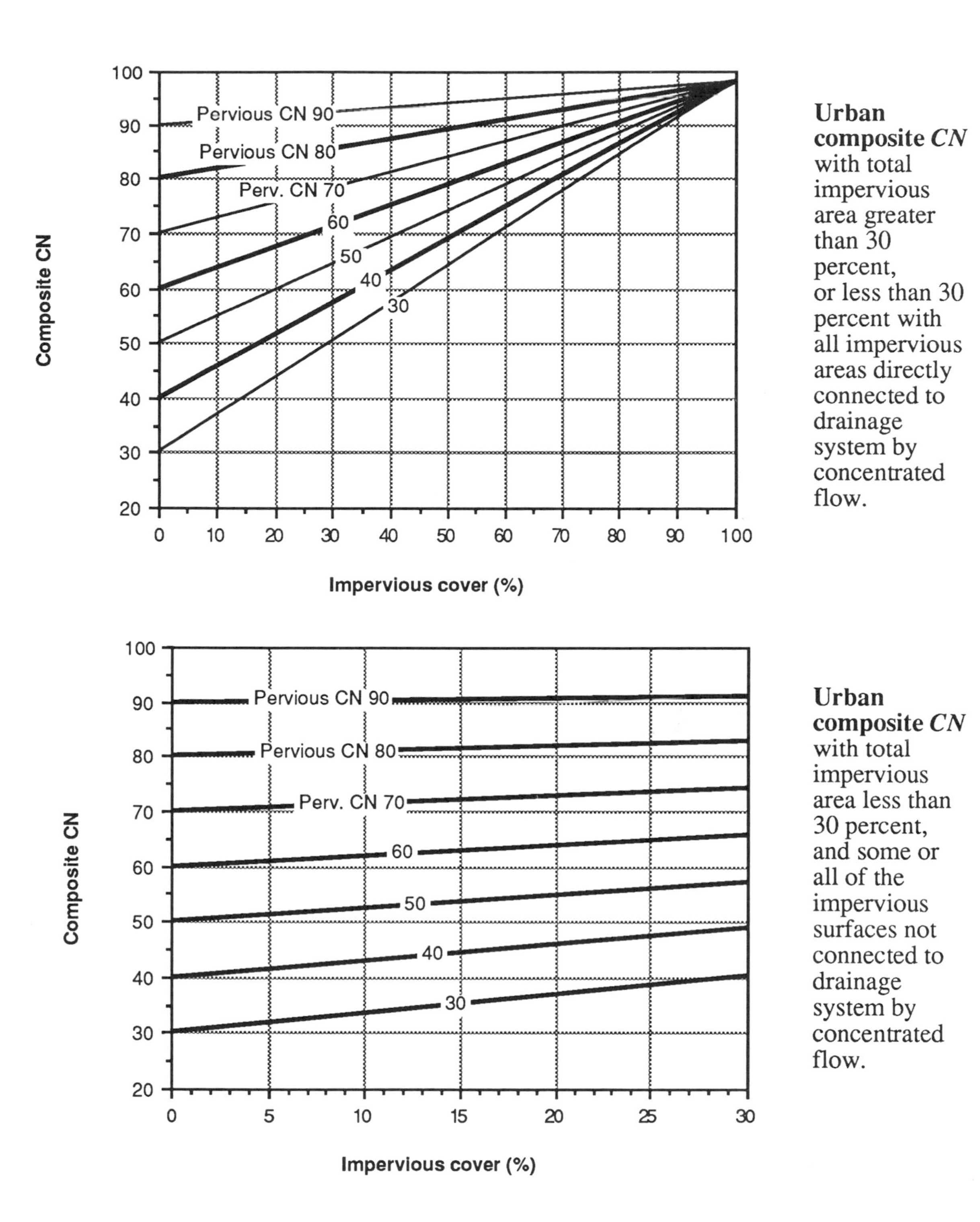

Urban composite *CN* with total impervious area greater than 30 percent, or less than 30 percent with all impervious areas directly connected to drainage system by concentrated flow.

Urban composite *CN* with total impervious area less than 30 percent, and some or all of the impervious surfaces not connected to drainage system by concentrated flow.

Charts on this page were derived from equations for Figures 2-3 and 2-4 given on page F-1 of U.S. Soil Conservation Service, 1986, *Urban Hydrology for Small Watersheds*, Technical Release 55.

Depth of runoff during the storm can now be found from the graph below. Enter the graph from the bottom, with the precipitation at a selected recurrence interval. Go vertically up to the line for your *CN*. From there move horizontally to the left, to read the depth of runoff *Dr*. As an alternative to the chart, you could use the equation on page 34, which is the basis for the chart.

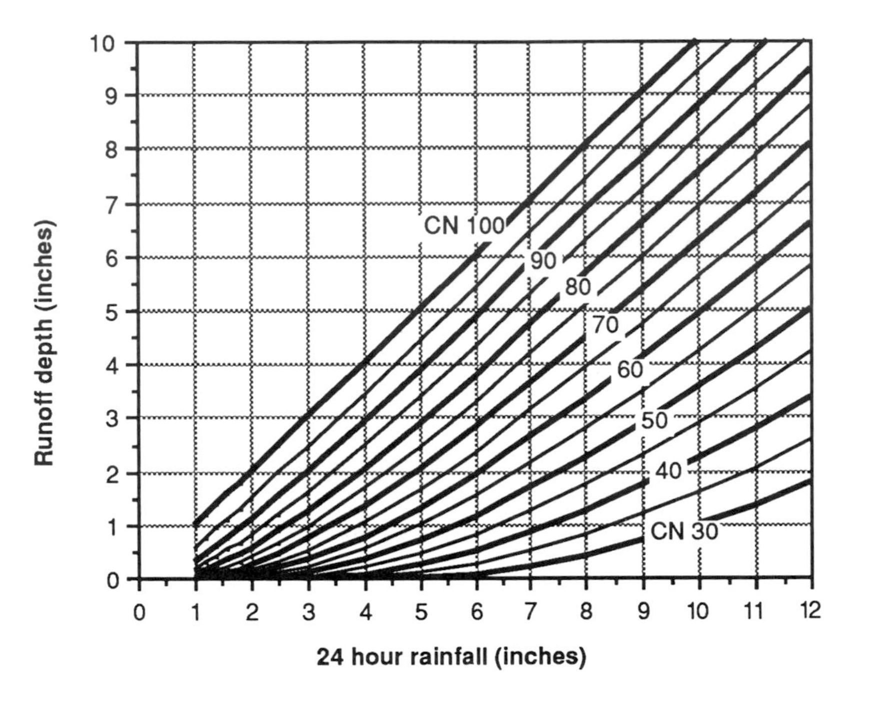

Derived from Equation 2-3 of U.S. Soil Conservation Service, 1986, Technical Release 55, *Urban Hydrology for Small Watersheds*, second edition.
Values in this chart are based on *Ia* of 0.2.

Volume of runoff *Qvol* is simply the depth times the watershed area:

$$Qvol = Dr\ Ad\ /\ 12$$

where,

Qvol = volume of runoff, ac.ft.;
Dr = depth of runoff, in.;
Ad = drainage area, acres; and
12 = conversion factor, in./ft.

To find *Qp* use the equation,

$$Qp = Qu\ Ad\ Dr\ Fp$$

where,

Qp = peak rate of flow, cfs;
Qu = unit peak discharge, cfs/ac/in;
Dr = depth of runoff, in.; and
Fp = factor for ponds and swamps outside the hydraulic length.

First find the ratio *Ia/P* (inital abstraction as a proportion of total precipitation) from the graph below.

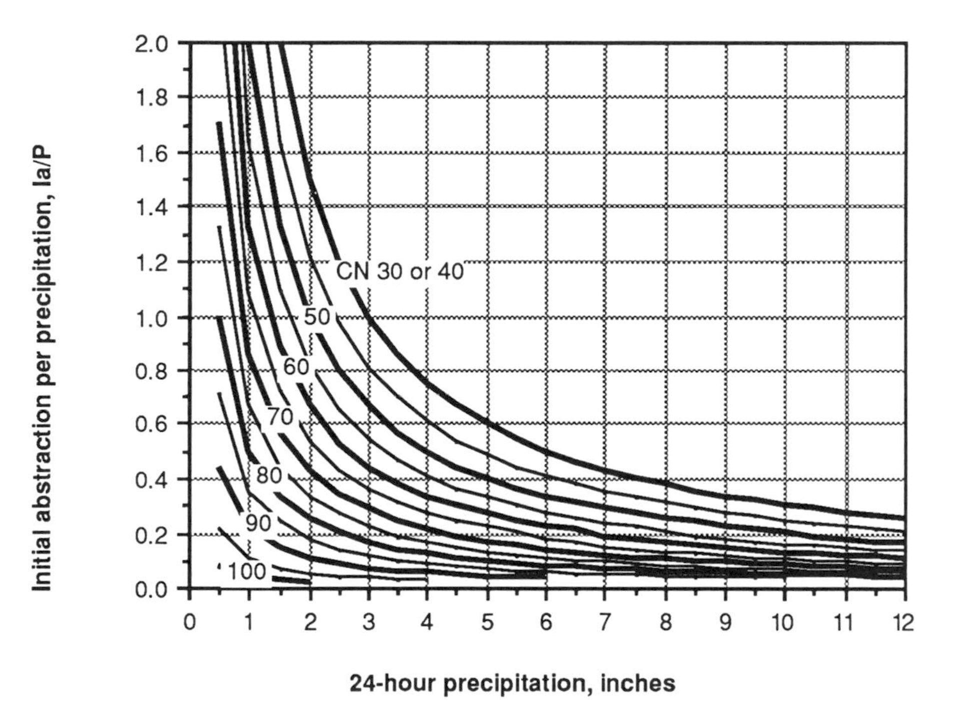

Derived from Table 4-1 of U.S. Soil Conservation Service, 1986, *Urban Hydrology for Small Watersheds*, Technical Release No. 55, second edition.

The intensity of rain within the 24-hour storm period is assumed by SCS to be distributed from hour to hour differently in each rainfall region of the country. The assumed distribution contains the intensity of any duration of storm equal to or less than 24 hours. For instance, the most intense hour within the 10-year, 24-hour storm has the intensity of the 10-year, 1-hour storm. The map on the next page shows which rainfall region your site is in.

Find the graph of unit peak discharge *Qu* for your rainfall distribution on page 45 or 46 according to your rainfall distribution type. Enter the appropriate graph from the bottom with your watershed's time of concentration. Go up to the line for your *Ia/P* ratio, then horizontally to the left to read unit peak discharge *Qu*, the peak flow per acre of watershed per inch of runoff volume.

SCS Rainfall Distributions

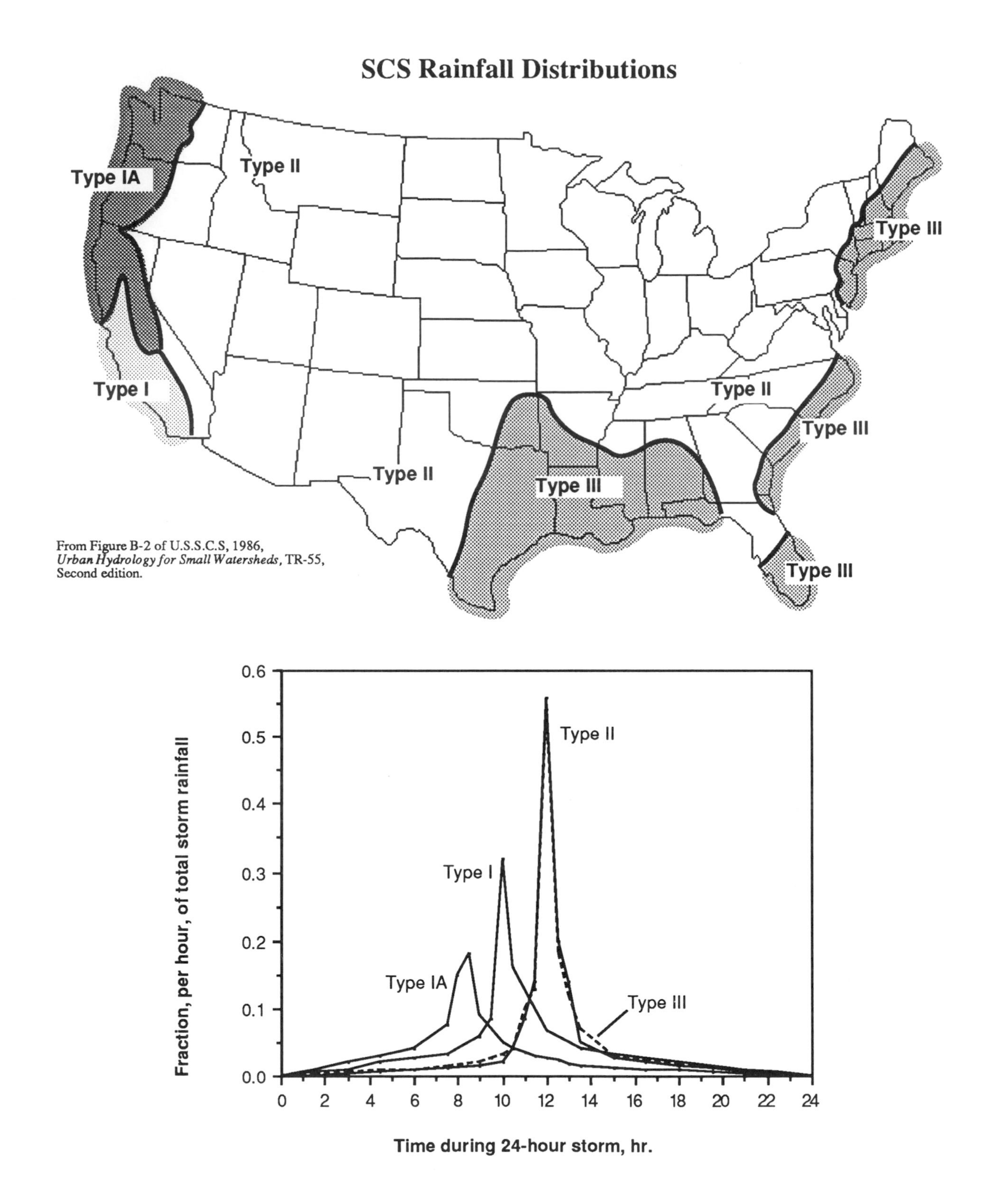

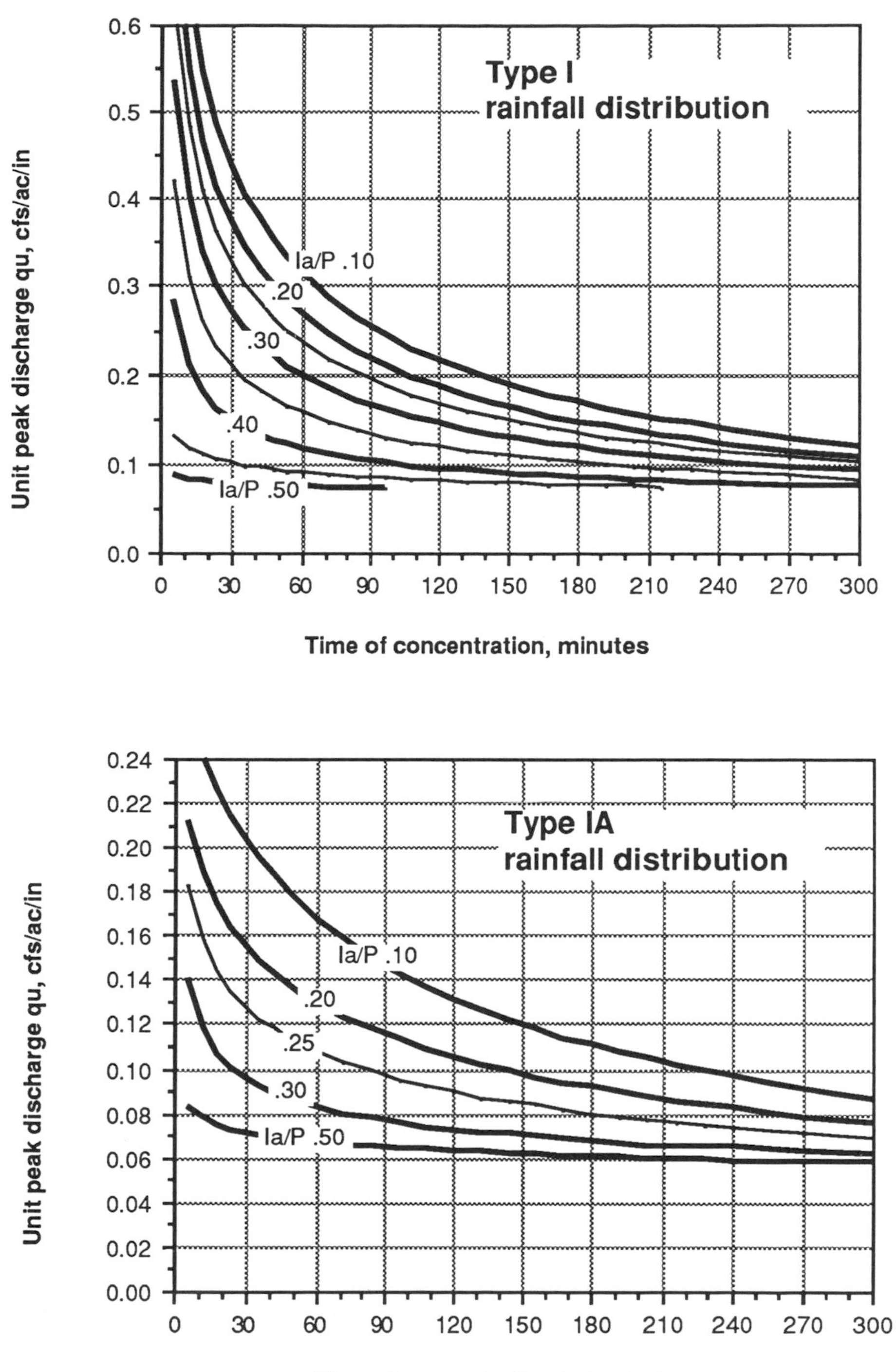

Charts on this page were derived from equation for Exhibit 4 given on page F-1 of U.S. Soil Conservation Service, 1986, *Urban Hydrology for Small Watersheds*, Technical Release 55, second edition.

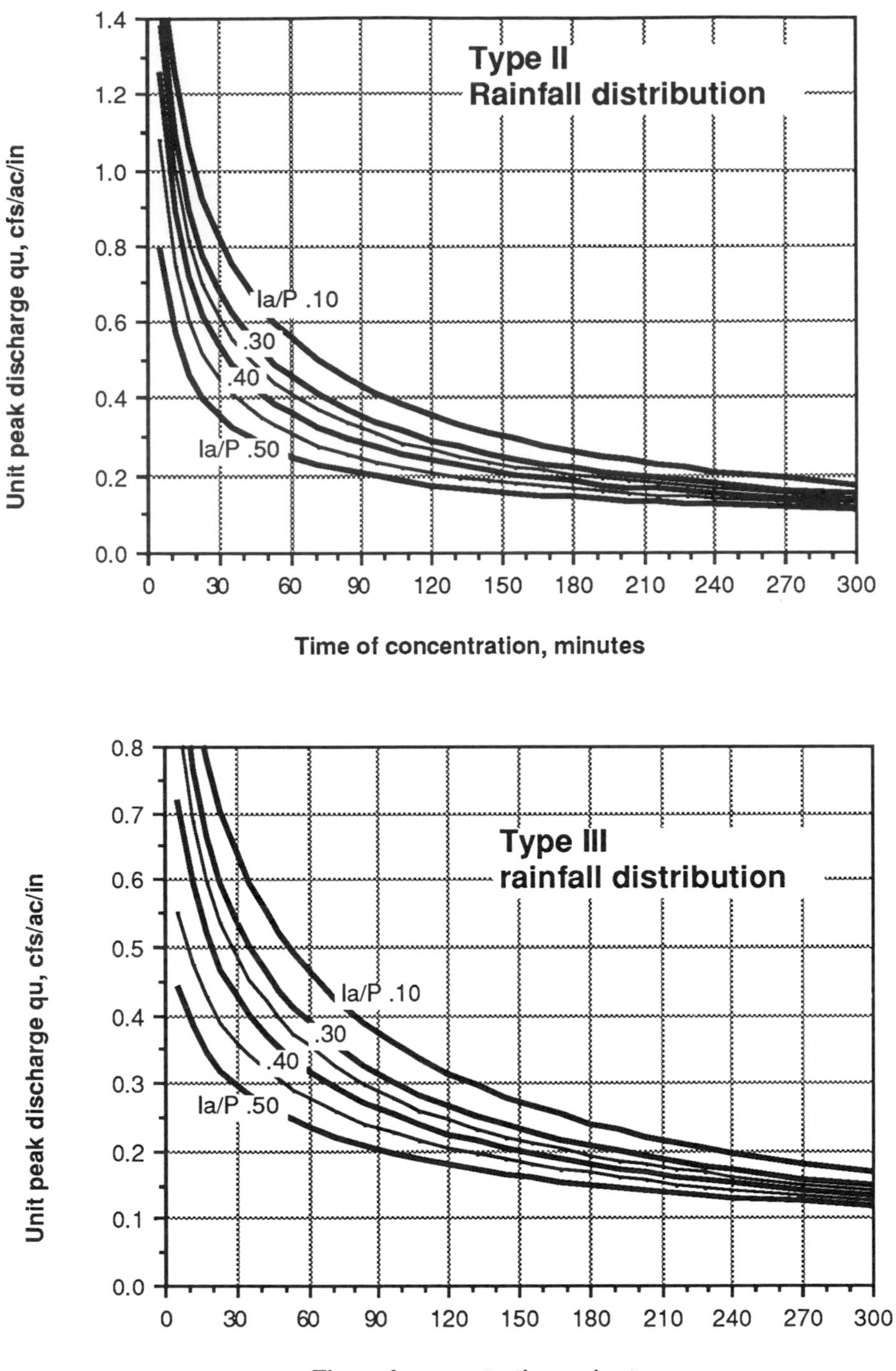

Charts on this page were derived from equation for Exhibit 4 given on page F-1 of U.S. Soil Conservation Service, 1986, *Urban Hydrology for Small Watersheds*, Technical Release 55, second edition.

The pond-and-swamp factor *Fp* is intended to take into account the slowing down of runoff as it passes through ponds or swamps. It applies to watersheds where pond and swamp areas are spread throughout the watershed and do not all lie along the hydraulic length, and so were not considered in the estimate of time of concentration. In watersheds that do not have such ponds or swamps, *Fp* is equal to one and can be disregarded. For watersheds that have such ponds or swamps, find *Fp* from the graph below and use it in the equation for *Qp* on page 43.

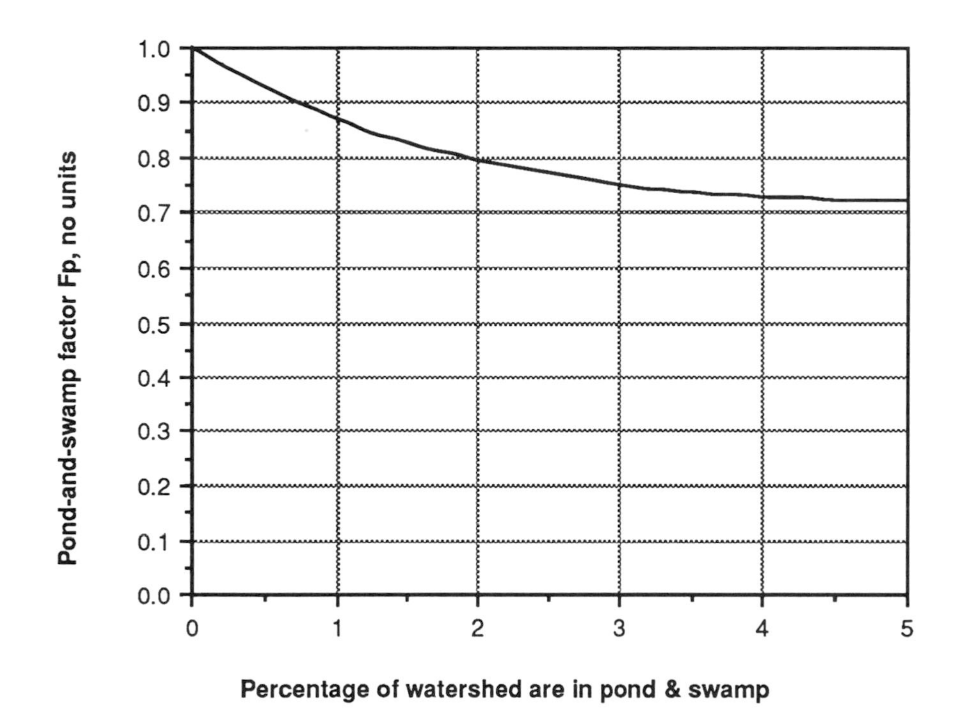

Finding peak rate of flow *Qp* is the complicated part of the SCS method. On a computer, the method involves repeated calculations to add up flows as intensity of rainfall and intensity of runoff vary over time. Charts like those in this book are based on many computer runs by SCS with their original equations.

The above manual method is referred to by SCS as the "graphical peak discharge" method, and specifies that it should be applied only to homogeneous individual watersheds. For watersheds that could be divided into subwatersheds with substantially different curve numbers or times of concentration due to land use, soil or slope, you should use either the "tabular hydrograph" method of *TR55* for manual calculations, or a computer program that can route flows from different subwatersheds together, such as the one that accompanies this book. Such approaches generate hydrographs, telling you how runoff varies from moment to moment during a storm, so that runoff from a number of subwatersheds can be tracked over time.

Summary of Process:
SCS Method

Depth of runoff:

1. Obtain 24-hour rainfall in inches, at a selected recurrence interval, from one of the maps or references on pages 35-36, or local data.

2. Obtain hydrologic soil group *HSG* from the local SCS soil survey that applies to your site, or by interpreting the definitions on page 37.

3. Obtain curve number *CN* (no units) from the tables on page 39 and 40. For a drainage area with more than one *HSG* or land use compute a weighted average *CN* for the area.

4. Find runoff depth *Dr* in inches from the graph on page 42.

Volume of runoff:

5. Estimate drainage area *Ad* in acres from a site map.

6. Compute volume of flow *Qvol* in ac.ft.:

$$Qvol = Dr\ Ad\ /\ 12$$

Peak rate of flow:

7. Obtain ratio of initial abstraction to precipitation *Ia/P* (no units) from the graph on page 43.

8. Find type of distribution of storm rainfall from the map on page 44.

9. Obtain time of concentration *Tc* of your drainage area, in minutes, from your estimate on page 25 or the appendix on *SCS Time of Concentration.*

10. Obtain unit peak discharge *Qu* (cfs/ac./in.) from the appropriate graph on page 45 or 46.

11. Obtain pond-and-swamp factor *Fp* (no units) from the graph on page 47. If there are no significant ponds or swamps in the drainage area, then *Fp* is equal to 1 and can be disregarded in the following step.

12. Compute peak rate of runoff *Qp* in cfs:

$$Qp = Qu\ Ad\ Dr\ Fp$$

Storm Runoff Estimation Exercise:
SCS Method

1. Before Development

		Site 1	***Site 2***
Runoff depth:			
24 hour rainfall (from page 35 or 36)	=	in	in
Hydrologic soil group *HSG* (from page 37)	=		
Curve number *CN* (from page 39 or 40)	=		
Runoff depth *Dr* (from page 42)	=	in	in
Runoff volume:			
Drainage area *Ad* (from site map)	=	ac	ac
Runoff volume *Qvol* = *Dr Ad* / 12	=	ac.ft.	ac.ft.
Peak rate of runoff:			
Ratio *Ia*/*P* (from page 43)	=		
Rainfall distribution type (from page 44)	=		
Time of concentration *Tc* (from page 25)	=		
Unit peak discharge *Qu* (from page 45 or 46)	=	cfs/ac/in	cfs/ac/in
Pond-and-swamp factor *Fp* (from page 47)	=		
Peak rate of flow *Qp* = *Qu Ad Dr Fp*	=	cfs	cfs

Storm Runoff Estimation Exercise:
SCS Method

1. After Development

		Site 1	***Site 2***
Runoff depth:			
24 hour rainfall (from page 35 or 36)	=	______ in	______ in
Hydrologic soil group *HSG* (from page 37)	=	______	______
Curve number *CN* (from page 39 or 40)	=	______	______
Runoff depth *Dr* (from page 42)	=	______ in	______ in
Runoff volume:			
Drainage area *Ad* (from site map)	=	______ ac	______ ac
Runoff volume *Qvol* = *Dr Ad* / 12	=	______ ac.ft.	______ ac.ft.
Peak rate of runoff:			
Ratio *Ia*/*P* (from page 43)	=	______	______
Rainfall distribution type (from page 44)	=	______	______
Time of concentration *Tc* (from page 25)	=	______	______
Unit peak discharge *Qu* (from page 45 or 46)	=	______ cfs/ac/in	______ cfs/ac/in
Pond-and-swamp factor *Fp* (from page 47)	=	______	______
Peak rate of flow *Qp* = *Qu Ad Dr Fp*	=	______ cfs	______ cfs

Runoff Estimation Exercise:
Discussion of Results

1. Which method, "rational" or SCS, suggests the larger peak rate of runoff (*Qp*) on each site? Why?

2. Would you select one runoff estimation method over another? On what grounds?

3. Outline the volume of one cubic foot with your hands. Do you think either of the runoff estimation methods is sufficiently precise to give you answers in fractions of a cubic foot? To what digit do you think you should round off your calculations of *Qvol* and *Qp*?

4. Which site has the larger volume (*Qvol*) and peak rate (*Qp*) of runoff *before* development? Why?

5. On which site does the volume of runoff (*Qvol*) increase by the greatest amount (ac.ft.) when the land is developed? Why? On which does it increase by the greatest proportion (*Qvol* after / *Qvol* before)? Why?

6. On which site does the peak rate of runoff (*Qp*) increase by the greatest amount (cfs) when the land is developed? Why? On which does it increase by the greatest proportion (*Qp* after / *Qp* before)? Why?

7. In terms of *Qp* and *Qvol*, does development cause the two sites to become hydrologically more similar, or more different? Why?

8. If you were planning a region's future land use for the purpose of preventing environmental damage to streams, on what types of soils, slopes and vegetation would you propose development with a high proportion of impervious cover? Why?

9. How could you alter a development's layout or program to reduce the total volume of runoff that is generated? Name several specific ways.

10. Given that a development will generate a certain total volume of runoff, how could you alter the layout so as to reduce the volume that reaches a given place in the drainage system? What effects would such alterations have on the runoff reaching other parts of the drainage system?

11. Rate of rainfall during a storm is expressed in units such as inches per hour or inches per 24 hours, but runoff is expressed in cfs. This difference in units of measurement often prevents us from appreciating relative magnitudes of runoff. From the data for the "rational" method, convert your site's peak intensity of rainfall in inches per hour to cfs using a conversion factor of 1.008 cfs per ac-in/hr:

$$\text{Intensity of rainfall in cfs} = 1.008\, I\, Ad$$

Then answer these questions: What is the difference between rate and volume of rainfall entering the watershed, and discharge leaving it, under existing and proposed conditions? What processes might occur in the watershed to create these differences?

12. Find the proportion of rainfall volume that becomes runoff volume by dividing the SCS runoff depth *Dr* by the 24-hour storm rainfall. What is the difference in the runoff/rainfall ratio between existing and proposed conditions? What processes might occur in the watershed to create such a difference? If some of the rain did not come out as storm runoff, where did it go? Will it ever reach the streams?

13. What is the total duration of your design-storm event? Divide the duration by the recurrence interval to find the proportion of total elapsed time occupied by the storm. Apply an appropriate conversion factor to make sure units are uniform, such as 8,766 hr/yr if storm duration is in hours and recurrence interval is in years. If you construct a pipe or detention basin for this storm, and then go visit the site on a randomly selected day of a randomly selected year, will you expect it to be full of flowing water? What *would* you expect to see there?

Summary and Commentary

You haven't designed anything yet. When you have estimated runoff, you have only *modeled* a *natural* process. You have made an estimate of what will happen *if* a certain rain falls on your site.

Accuracy of runoff models has been the subject of long arguments. Many of those arguments have in fact been unresolvable, because they have taken place without the benefit of actual measured on-site runoff data. The SCS model has often been believed to be inherently more accurate than the "rational" because it is more "sophisticated," in the sense that it is more complex and takes a greater number of factors explicitly into account. But *sophistication* does not equal *accuracy.* The SCS method has also been promoted as embodying a superior *theory* of the mechanisms by which runoff is generated through soils, impervious surfaces, etc. But a superior theory (if it is really there) does not in itself create superior accuracy, any more than superior sophistication does.

Accuracy is a positive relationship between the *results* of a model — the estimated volume and rate of runoff — and what *actually* happens on real drainage areas, during real rain storms. A high level of accuracy means that the results of a model closely approximate *real* flows.

Accurate results can be *assured* only when the runoff model you are using has been calibrated to actual local conditions. Calibrating a model requires acquisition of local data. Several precipitation and stream gauging stations need to be set up for a period of at least 3 to 5 years in a watershed with stable land use. Streamflow and rainfall data from storm events would then be collected. Using these data, a selected model can be calibrated or manipulated to yield estimates of runoff similar to what has been observed in the field. The calibrated model can then be used for runoff estimates in all nearby areas that have hydrologic characteristics similar to the sites where gauging took place.

The computer disk that accompanies this book has not been calibrated to any particular local conditions. It contains only the general SCS-method equations.

The SCS model is more amenable to such calibration than the "rational," because it has more variables describing drainage area and precipitation characteristics. Each of these variables can be changed or calibrated to describe local conditions relatively precisely, and thus to match locally observed data relatively closely.

Too frequently we have to design facilities that are going to get built right away, in locales that have *not* been monitored even for three years. In these cases models must be used to make estimates of natural processes based on general knowledge, with the understanding that site-specific accuracy is not a definable issue.

Which model to choose for a specific project is often a matter of many considerations other than accuracy. The SCS method tends to be more *consistent* in its application. Depending on local conditions, one method or the other might be more *conservative*, in the sense that its use usually ends up requiring facilities with greater capacities. Local agencies that must review development plans often *accept* only one particular method.

Snowfall and snowmelt are major components of precipitation and runoff in some regions, such as mountainous parts of the Pacific Northwest. In those places runoff models can be used which take into account snowfall and snowmelt as well as rainfall. The methods discussed in this chapter do not take account of snowfall or snowmelt. If you work in such a region you can find out about locally preferred models by referring to hydrologic guides published by local agencies.

Chapter 3

Water Balance Estimation

A water balance of a stormwater basin is a summary of all its inflows and outflows over a period of time. It is a way to take into account the low-level background flows, in contrast to the abrupt, infrequent pulses of runoff reflected in design storm.

Like storm runoff estimation, estimation of a water balance defines the volumes and rates of flows that we have to deal with in design. It is a modeling of a natural process. It is a foundation of many design applications, supplementing estimation of storm runoff.

Consider a design storm with a 10-year recurrence interval and 24 hours duration. Such a storm occupies a minute fraction of one percent of the total elapsed time during its recurrence interval.

During the intervening years there are ongoing flows of precipitation, run-off, evaporation, transpiration and infiltration. At the average U.S. rainfall of 30 inches per year, 25 feet of water pass by while we wait for each occurrence of a storm which is only a few inches in magnitude.

Background flows represent a resource. If you want a permanent pond for water-quality control, wildlife or community amenity, it must be adequately supported by the background flows during the dry months of average years. If your environmental objectives call for a perennially flowing stream downdrainage, or recharge of an underlying aquifer, their support depends on the regularly occurring low-level flows in which most of the total annual flow is contained.

Background flows also represent a hazard. They can accumulate in an infiltration basin where standing water may not be desired. By filling part of the basin, they could reduce the capacity remaining to capture a design storm, or make a basin that was supposed to be dry turf into a mudpond.

A water balance is needed for infiltration basins and permanent pools, where long-term flows accumulate below the elevation of any surface outlet. It can tell the designer about everyday presence of standing water, moisture endowments to vegetation and wildlife, or long-term disposition of water from the basin into the environment. Water-balance analysis can supplement, not replace, design-storm analysis.

For application to design, a water balance must be constructed that will answer the specific design question you have. This chapter introduces the concept of the water balance, and sets up the necessary types of data so that they can be used when the time for design comes.

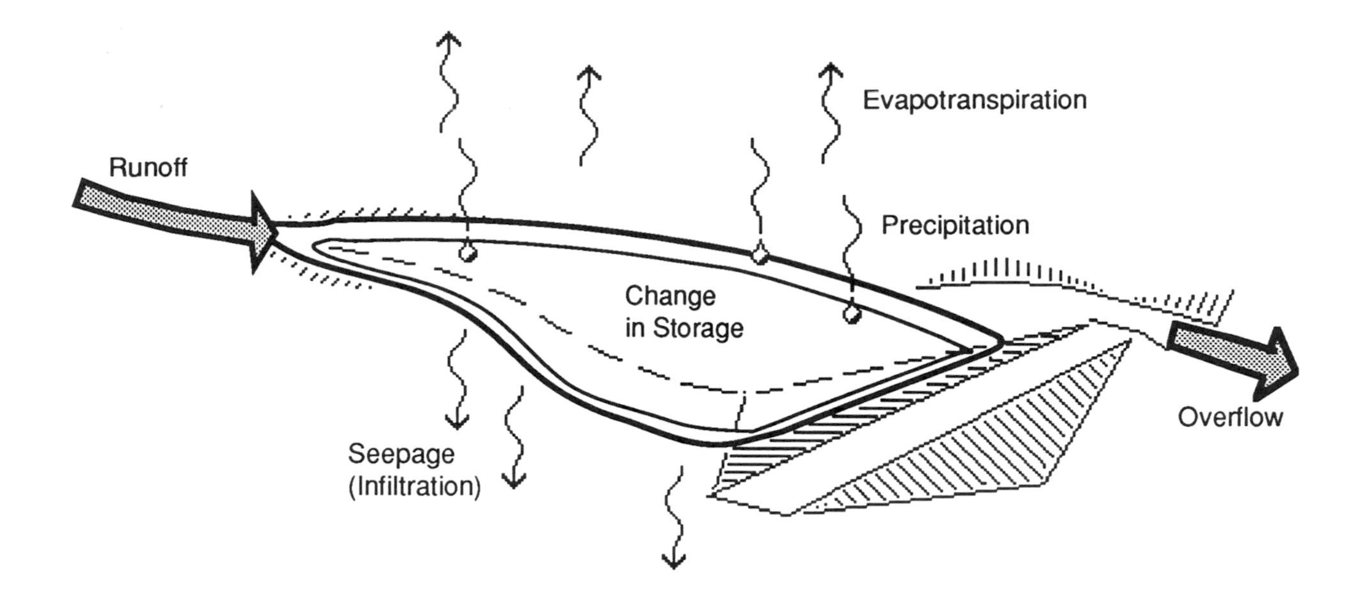

The basic water-balance equation for a stormwater basin is the same as for any watershed, water body or reservoir. It is the change-of-volume equation that was introduced on page 8. During any period of time,

Change in storage = Inflows - Outflows

The equation says simply that, whatever the inflows and outflows might be, any difference between them is taken up by the amount of water remaining in the basin. When there is more inflow than outflow, the amount of water in storage increases. When there is more outflow, the storage declines. This is why it is called a water "balance." By definition, the two sides of the equation must be equal, or balanced.

The potential inflows to a stormwater basin are runoff from the watershed area that drains to it (the "catchment" area), and precipitation directly upon the basin.

The potential outflows from a stormwater basin are infiltration, surface overflow, and either evapotranspiration or open-water evaporation depending on whether the basin is wet or dry.

Water-balance data for designing basins must reflect long-term averages, or a recurrence interval accommodating an acceptable level of risk. As in other kinds of hydraulic design, a longer recurrence interval might need to be traded off against greater construction cost.

The tables on the following pages show monthly long-term average climatic data for U.S. cities, which can be combined with site-specific watershed characteristics to yield components of the water balance.

Water Balance Data:
Long-term average

	Jan	Feb	Mar	Apr	May	Jun	Jul	Aug	Sep	Oct	Nov	Dec	Annual
ALABAMA													
Birmingham, Ala.													
Precipitation P, ft.	0.46	0.47	0.50	0.38	0.29	0.32	0.41	0.34	0.26	0.25	0.33	0.43	4.45
Turf evapotranspiration Et, ft.	0.06	0.07	0.15	0.28	0.44	0.58	0.63	0.58	0.42	0.26	0.11	0.06	3.66
Lake evaporation El, ft.	0.16	0.20	0.30	0.38	0.49	0.51	0.46	0.45	0.38	0.28	0.16	0.14	3.90
ARIZONA													
Flagstaff, Ariz.													
Precipitation P, ft.	0.10	0.10	0.09	0.07	0.04	0.04	0.15	0.19	0.11	0.09	0.06	0.10	1.16
Turf evapotranspiration Et, ft.	0.02	0.04	0.09	0.17	0.29	0.43	0.52	0.46	0.32	0.17	0.06	0.03	2.59
Lake evaporation El, ft.	0.13	0.16	0.25	0.41	0.58	0.69	0.59	0.51	0.43	0.29	0.17	0.11	4.31
Phoenix, Ariz.													
Precipitation P, ft.	0.08	0.08	0.06	0.03	0.01	0.01	0.09	0.13	0.07	0.05	0.05	0.08	0.75
Turf evapotranspiration Et, ft.	0.08	0.11	0.21	0.35	0.53	0.69	0.84	0.75	0.56	0.34	0.16	0.10	4.73
Lake evaporation El, ft.	0.23	0.30	0.46	0.65	0.84	0.94	0.91	0.79	0.65	0.51	0.29	0.21	6.77
Tucson, Ariz.													
Precipitation P, ft.	0.08	0.08	0.06	0.03	0.01	0.04	0.23	0.26	0.11	0.07	0.05	0.08	1.12
Turf evapotranspiration Et, ft.	0.06	0.09	0.17	0.29	0.44	0.60	0.68	0.60	0.46	0.29	0.13	0.08	3.89
Lake evaporation El, ft.	0.30	0.35	0.54	0.72	0.91	0.99	0.83	0.72	0.70	0.57	0.38	0.29	7.29
ARKANSAS													
Little Rock, Ark.													
Precipitation P, ft.	0.41	0.37	0.42	0.45	0.48	0.34	0.31	0.26	0.28	0.27	0.36	0.35	4.30
Turf evapotranspiration Et, ft.	0.05	0.07	0.15	0.29	0.45	0.60	0.68	0.62	0.43	0.27	0.11	0.06	3.81
Lake evaporation El, ft.	0.16	0.19	0.31	0.41	0.50	0.59	0.59	0.54	0.39	0.31	0.17	0.15	4.30
CALIFORNIA													
Fresno, Cal.													
Precipitation P, ft.	0.30	0.30	0.25	0.16	0.07	0.02	0.00	0.00	0.02	0.08	0.14	0.30	1.63
Turf evapotranspiration Et, ft.	0.05	0.08	0.15	0.24	0.36	0.47	0.60	0.53	0.39	0.24	0.12	0.06	3.30
Lake evaporation El, ft.	0.09	0.18	0.33	0.48	0.68	0.80	0.86	0.74	0.57	0.37	0.14	0.06	5.29
Los Angeles, Cal.													
Precipitation P, ft.	0.27	0.30	0.22	0.12	0.03	0.01	0.01	0.01	0.02	0.06	0.11	0.26	1.41
Turf evapotranspiration Et, ft.	0.08	0.10	0.18	0.25	0.34	0.41	0.51	0.48	0.38	0.26	0.15	0.10	3.25
Lake evaporation El, ft.	0.23	0.25	0.32	0.37	0.41	0.41	0.52	0.48	0.39	0.30	0.23	0.21	4.13
Sacramento, Cal.													
Precipitation P, ft.	0.57	0.52	0.40	0.24	0.14	0.06	0.01	0.01	0.04	0.17	0.30	0.54	2.99
Turf evapotranspiration Et, ft.	0.04	0.06	0.12	0.20	0.31	0.42	0.54	0.47	0.34	0.21	0.09	0.05	2.87
Lake evaporation El, ft.	0.08	0.16	0.29	0.43	0.57	0.74	0.81	0.71	0.55	0.34	0.14	0.06	4.90

	Jan	Feb	Mar	Apr	May	Jun	Jul	Aug	Sep	Oct	Nov	Dec	Annual
San Francisco, Cal.													
Precipitation P, ft.	0.34	0.31	0.24	0.13	0.05	0.01	0.00	0.00	0.01	0.07	0.14	0.33	1.65
Turf evapotranspiration Et, ft.	0.07	0.09	0.16	0.23	0.31	0.37	0.42	0.38	0.32	0.23	0.13	0.08	2.78
Lake evaporation El, ft.	0.12	0.20	0.28	0.38	0.44	0.49	0.47	0.42	0.39	0.28	0.16	0.11	3.74
COLORADO													
Denver, Colo.													
Precipitation P, ft.	0.04	0.05	0.09	0.16	0.21	0.14	0.15	0.13	0.10	0.08	0.05	0.04	1.23
Turf evapotranspiration Et, ft.	0.00	0.00	0.00	0.10	0.19	0.30	0.37	0.32	0.20	0.10	0.00	0.00	1.58
Lake evaporation El, ft.	0.16	0.17	0.26	0.41	0.49	0.60	0.67	0.60	0.44	0.34	0.19	0.16	4.49
Grand Junction-Durango, Colo.													
Precipitation P, ft.	0.11	0.11	0.12	0.12	0.10	0.08	0.11	0.14	0.11	0.11	0.08	0.10	1.32
Turf evapotranspiration Et, ft.	0.00	0.00	0.04	0.12	0.22	0.33	0.43	0.37	0.23	0.12	0.03	0.00	1.90
Lake evaporation El, ft.	0.09	0.14	0.29	0.44	0.65	0.79	0.86	0.75	0.57	0.40	0.15	0.08	5.21
CONNECTICUT													
Hartford, Conn.													
Precipitation P, ft.	0.32	0.25	0.36	0.33	0.32	0.30	0.33	0.37	0.34	0.28	0.37	0.32	3.90
Turf evapotranspiration Et, ft.	0.00	0.00	0.06	0.15	0.30	0.44	0.53	0.46	0.30	0.16	0.06	0.00	2.46
Lake evaporation El, ft.	0.11	0.12	0.21	0.34	0.43	0.48	0.49	0.43	0.29	0.21	0.13	0.09	3.34
DISTRICT OF COLUMBIA													
Washington, D.C.													
Precipitation P, ft.	0.23	0.20	0.28	0.28	0.34	0.31	0.36	0.38	0.29	0.29	0.24	0.23	3.42
Turf evapotranspiration Et, ft.	0.03	0.04	0.09	0.21	0.38	0.51	0.57	0.51	0.34	0.20	0.08	0.03	2.99
Lake evaporation El, ft.	0.17	0.17	0.28	0.39	0.47	0.55	0.56	0.50	0.37	0.24	0.20	0.15	4.05
FLORIDA													
Jacksonville-Daytona Beach, Fla.													
Precipitation P, ft.	0.21	0.27	0.33	0.29	0.28	0.51	0.63	0.57	0.55	0.37	0.16	0.21	4.39
Turf evapotranspiration Et, ft.	0.11	0.13	0.25	0.37	0.53	0.62	0.66	0.63	0.50	0.36	0.20	0.13	4.49
Lake evaporation El, ft.	0.22	0.26	0.39	0.48	0.55	0.51	0.48	0.43	0.38	0.35	0.24	0.20	4.48
Tampa, Fla.													
Precipitation P, ft.	0.17	0.22	0.30	0.28	0.29	0.61	0.68	0.64	0.65	0.34	0.15	0.15	4.49
Turf evapotranspiration Et, ft.	0.14	0.17	0.29	0.41	0.55	0.62	0.66	0.63	0.52	0.39	0.24	0.16	4.79
Lake evaporation El, ft.	0.23	0.28	0.41	0.49	0.61	0.55	0.51	0.46	0.42	0.37	0.26	0.22	4.82
GEORGIA													
Atlanta, Ga.													
Precipitation P, ft.	0.42	0.42	0.46	0.39	0.30	0.31	0.42	0.33	0.27	0.25	0.27	0.41	4.28
Turf evapotranspiration Et, ft.	0.06	0.07	0.15	0.27	0.44	0.56	0.61	0.56	0.41	0.25	0.11	0.06	3.55
Lake evaporation El, ft.	0.19	0.21	0.31	0.41	0.51	0.52	0.48	0.47	0.37	0.32	0.20	0.17	4.16
IDAHO													
Boise, Id.													
Precipitation P, ft.	0.13	0.11	0.11	0.09	0.10	0.08	0.02	0.01	0.04	0.08	0.11	0.12	1.00
Turf evapotranspiration Et, ft.	0.00	0.03	0.09	0.19	0.32	0.42	0.58	0.49	0.30	0.15	0.05	0.02	2.63
Lake evaporation El, ft.	0.07	0.11	0.25	0.40	0.51	0.63	0.83	0.71	0.48	0.28	0.11	0.06	4.44

	Jan	Feb	Mar	Apr	May	Jun	Jul	Aug	Sep	Oct	Nov	Dec	Annual
ILLINOIS													
Chicago, Ill.													
Precipitation *P*, ft.	0.15	0.13	0.22	0.27	0.31	0.34	0.28	0.28	0.24	0.22	0.17	0.16	2.79
Turf evapotranspiration *Et*, ft.	0.00	0.00	0.06	0.17	0.33	0.49	0.58	0.51	0.32	0.17	0.05	0.00	2.68
Lake evaporation *El*, ft.	0.11	0.13	0.21	0.31	0.47	0.58	0.59	0.50	0.35	0.28	0.16	0.10	3.78
INDIANA													
Indianapolis, Ind.													
Precipitation *P*, ft.	0.26	0.20	0.28	0.31	0.36	0.37	0.29	0.25	0.28	0.21	0.25	0.21	3.28
Turf evapotranspiration *Et*, ft.	0.00	0.02	0.08	0.19	0.35	0.52	0.59	0.52	0.34	0.19	0.06	0.02	2.90
Lake evaporation *El*, ft.	0.11	0.13	0.20	0.34	0.41	0.52	0.52	0.47	0.35	0.25	0.14	0.09	3.54
IOWA													
Des Moines, Iowa													
Precipitation *P*, ft.	0.10	0.09	0.17	0.22	0.35	0.41	0.30	0.31	0.26	0.17	0.15	0.09	2.61
Turf evapotranspiration *Et*, ft.	0.00	0.00	0.05	0.17	0.34	0.49	0.59	0.50	0.32	0.17	0.04	0.00	2.67
Lake evaporation *El*, ft.	0.08	0.09	0.16	0.32	0.44	0.55	0.58	0.49	0.32	0.28	0.14	0.09	3.52
KANSAS													
Wichita, Kans.													
Precipitation *P*, ft.	0.06	0.09	0.12	0.20	0.34	0.33	0.25	0.23	0.21	0.16	0.09	0.07	2.16
Turf evapotranspiration *Et*, ft.	0.02	0.04	0.10	0.24	0.41	0.59	0.70	0.63	0.41	0.23	0.08	0.03	3.49
Lake evaporation *El*, ft.	0.14	0.18	0.33	0.47	0.52	0.65	0.73	0.68	0.46	0.34	0.21	0.15	4.86
KENTUCKY													
Louisville, Ky.													
Precipitation *P*, ft.	0.43	0.32	0.42	0.34	0.33	0.36	0.35	0.29	0.25	0.21	0.30	0.32	3.91
Turf evapotranspiration *Et*, ft.	0.03	0.05	0.11	0.24	0.40	0.54	0.60	0.54	0.38	0.22	0.09	0.04	3.24
Lake evaporation *El*, ft.	0.12	0.14	0.22	0.36	0.42	0.50	0.52	0.49	0.35	0.25	0.15	0.12	3.65
LOUISIANA													
New Orleans, La.													
Precipitation *P*, ft.	0.36	0.38	0.43	0.41	0.39	0.46	0.64	0.56	0.55	0.26	0.32	0.42	5.20
Turf evapotranspiration *Et*, ft.	0.11	0.13	0.24	0.37	0.53	0.64	0.67	0.64	0.50	0.35	0.19	0.12	4.51
Lake evaporation *El*, ft.	0.17	0.23	0.34	0.41	0.43	0.49	0.44	0.41	0.39	0.34	0.23	0.18	4.05
MAINE													
Portland, Me.													
Precipitation *P*, ft.	0.35	0.30	0.33	0.30	0.28	0.27	0.26	0.23	0.31	0.31	0.39	0.33	3.67
Turf evapotranspiration *Et*, ft.	0.00	0.00	0.04	0.11	0.23	0.35	0.44	0.39	0.25	0.13	0.05	0.00	2.00
Lake evaporation *El*, ft.	0.08	0.10	0.15	0.25	0.33	0.41	0.45	0.38	0.24	0.17	0.09	0.09	2.75
MARYLAND													
Baltimore, Md.													
Precipitation *P*, ft.	0.27	0.23	0.32	0.30	0.35	0.32	0.36	0.38	0.31	0.28	0.27	0.27	3.65
Turf evapotranspiration *Et*, ft.	0.02	0.03	0.09	0.20	0.37	0.50	0.58	0.52	0.34	0.19	0.08	0.03	2.96
Lake evaporation *El*, ft.	0.14	0.16	0.27	0.36	0.44	0.52	0.54	0.46	0.35	0.25	0.17	0.13	3.79

	Jan	Feb	Mar	Apr	May	Jun	Jul	Aug	Sep	Oct	Nov	Dec	Annual
<u>MASSACHUSETTS</u>													
Boston, Mass.													
Precipitation *P*, ft.	0.34	0.28	0.35	0.32	0.27	0.26	0.24	0.32	0.30	0.28	0.34	0.31	3.61
Turf evapotranspiration *Et*, ft.	0.00	0.00	0.06	0.14	0.28	0.41	0.51	0.46	0.30	0.17	0.07	0.02	2.43
Lake evaporation *El*, ft.	0.18	0.17	0.26	0.36	0.45	0.55	0.60	0.51	0.38	0.29	0.20	0.17	4.10
<u>MICHIGAN</u>													
Detroit, Mich.													
Precipitation *P*, ft.	0.16	0.16	0.19	0.24	0.28	0.27	0.23	0.24	0.21	0.22	0.18	0.16	2.56
Turf evapotranspiration *Et*, ft.	0.00	0.00	0.05	0.15	0.30	0.46	0.55	0.47	0.31	0.16	0.05	0.00	2.50
Lake evaporation *El*, ft.	0.09	0.11	0.18	0.29	0.42	0.54	0.59	0.48	0.34	0.24	0.13	0.10	3.51
Marquette-Houghton, Mich.													
Precipitation *P*, ft.	0.14	0.12	0.15	0.20	0.27	0.35	0.30	0.31	0.28	0.19	0.22	0.14	2.67
Turf evapotranspiration *Et*, ft.	0.00	0.00	0.00	0.10	0.23	0.37	0.45	0.39	0.22	0.12	0.02	0.00	1.89
Lake evaporation *El*, ft.	0.08	0.08	0.12	0.22	0.34	0.40	0.48	0.38	0.24	0.17	0.08	0.08	2.65
Muskegon-Grand Rap., Mich.													
Precipitation *P*, ft.	0.19	0.16	0.16	0.22	0.25	0.26	0.22	0.24	0.26	0.22	0.24	0.17	2.60
Turf evapotranspiration *Et*, ft.	0.00	0.00	0.03	0.13	0.27	0.42	0.50	0.44	0.27	0.14	0.04	0.00	2.26
Lake evaporation *El*, ft.	0.08	0.09	0.14	0.27	0.39	0.49	0.53	0.43	0.25	0.18	0.08	0.06	2.99
<u>MINNESOTA</u>													
Minneapolis-St. Paul, Minn.													
Precipitation *P*, ft.	0.08	0.07	0.15	0.19	0.32	0.39	0.29	0.33	0.26	0.15	0.14	0.08	2.45
Turf evapotranspiration *Et*, ft.	0.00	0.00	0.00	0.14	0.31	0.46	0.55	0.48	0.28	0.14	0.03	0.00	2.40
Lake evaporation *El*, ft.	0.08	0.08	0.14	0.29	0.42	0.52	0.55	0.46	0.29	0.22	0.10	0.07	3.23
<u>MISSISSIPPI</u>													
Jackson-Canton, Miss.													
Precipitation *P*, ft.	0.46	0.43	0.50	0.39	0.34	0.31	0.43	0.27	0.23	0.19	0.34	0.43	4.34
Turf evapotranspiration *Et*, ft.	0.07	0.09	0.18	0.31	0.48	0.61	0.67	0.62	0.45	0.28	0.13	0.08	3.97
Lake evaporation *El*, ft.	0.18	0.21	0.32	0.39	0.49	0.54	0.50	0.46	0.36	0.29	0.18	0.13	4.05
<u>MISSOURI</u>													
Kansas City, Mo.													
Precipitation *P*, ft.	0.12	0.11	0.20	0.27	0.37	0.46	0.29	0.34	0.29	0.22	0.15	0.12	2.96
Turf evapotranspiration *Et*, ft.	0.00	0.02	0.08	0.21	0.38	0.55	0.65	0.57	0.37	0.21	0.06	0.02	3.12
Lake evaporation *El*, ft.	0.15	0.18	0.29	0.45	0.51	0.60	0.70	0.57	0.35	0.35	0.20	0.15	4.50
St. Louis, Mo.													
Precipitation *P*, ft.	0.15	0.15	0.24	0.30	0.34	0.39	0.30	0.29	0.26	0.25	0.20	0.15	3.03
Turf evapotranspiration *Et*, ft.	0.00	0.03	0.09	0.21	0.38	0.55	0.63	0.56	0.37	0.21	0.07	0.02	3.13
Lake evaporation *El*, ft.	0.13	0.16	0.26	0.40	0.47	0.57	0.58	0.50	0.37	0.29	0.17	0.12	4.01
<u>MONTANA</u>													
Helena, Mont.													
Precipitation *P*, ft.	0.05	0.05	0.07	0.09	0.17	0.25	0.11	0.10	0.10	0.08	0.06	0.05	1.18
Turf evapotranspiration *Et*, ft.	0.00	0.00	0.00	0.11	0.24	0.33	0.47	0.39	0.22	0.11	0.03	0.00	1.91
Lake evaporation *El*, ft.	0.07	0.10	0.18	0.28	0.43	0.48	0.64	0.53	0.31	0.22	0.09	0.08	3.39

	Jan	Feb	Mar	Apr	May	Jun	Jul	Aug	Sep	Oct	Nov	Dec	Annual
NEBRASKA													
Omaha, Neb.													
Precipitation *P*, ft.	0.07	0.08	0.13	0.20	0.31	0.38	0.26	0.29	0.22	0.12	0.09	0.06	2.21
Turf evapotranspiration *Et*, ft.	0.00	0.00	0.06	0.18	0.35	0.52	0.63	0.55	0.34	0.19	0.05	0.00	2.89
Lake evaporation *El*, ft.	0.11	0.13	0.23	0.37	0.51	0.58	0.64	0.53	0.34	0.30	0.14	0.10	3.97
NEVADA													
Las Vegas, Nev.													
Precipitation *P*, ft.	0.05	0.05	0.04	0.03	0.01	0.01	0.04	0.04	0.03	0.03	0.03	0.05	0.42
Turf evapotranspiration *Et*, ft.	0.05	0.08	0.17	0.30	0.46	0.62	0.77	0.68	0.48	0.27	0.12	0.06	4.06
Lake evaporation *El*, ft.	0.26	0.33	0.54	0.75	0.93	1.09	1.10	0.96	0.79	0.54	0.30	0.22	7.80
NEW HAMPSHIRE													
Concord, N.H.													
Precipitation *P*, ft.	0.29	0.22	0.29	0.29	0.29	0.30	0.31	0.26	0.32	0.25	0.32	0.28	3.43
Turf evapotranspiration *Et*, ft.	0.00	0.00	0.04	0.12	0.27	0.41	0.49	0.42	0.26	0.13	0.05	0.00	2.19
Lake evaporation *El*, ft.	0.07	0.09	0.16	0.25	0.34	0.41	0.45	0.35	0.22	0.16	0.09	0.08	2.67
NEW JERSEY													
Atlantic City, N.J.													
Precipitation *P*, ft.	0.29	0.26	0.33	0.28	0.27	0.26	0.32	0.41	0.31	0.27	0.30	0.28	3.58
Turf evapotranspiration *Et*, ft.	0.03	0.03	0.08	0.17	0.33	0.47	0.57	0.51	0.35	0.21	0.09	0.04	2.89
Lake evaporation *El*, ft.	0.15	0.16	0.24	0.36	0.42	0.47	0.50	0.44	0.32	0.25	0.18	0.14	3.64
NEW MEXICO													
Albuquerque, N.M.													
Precipitation *P*, ft.	0.03	0.04	0.03	0.04	0.05	0.06	0.12	0.13	0.10	0.07	0.03	0.04	0.74
Turf evapotranspiration *Et*, ft.	0.03	0.05	0.12	0.23	0.38	0.53	0.60	0.54	0.37	0.21	0.08	0.04	3.17
Lake evaporation *El*, ft.	0.19	0.24	0.42	0.60	0.77	0.87	0.78	0.68	0.56	0.43	0.23	0.17	5.94
NEW YORK													
Albany, N.Y.													
Precipitation *P*, ft.	0.25	0.22	0.28	0.30	0.32	0.32	0.36	0.32	0.33	0.27	0.30	0.27	3.54
Turf evapotranspiration *Et*, ft.	0.00	0.00	0.05	0.15	0.31	0.45	0.53	0.46	0.30	0.16	0.06	0.00	2.48
Lake evaporation *El*, ft.	0.09	0.10	0.17	0.29	0.39	0.46	0.51	0.40	0.27	0.18	0.10	0.08	3.04
Buffalo, N.Y.													
Precipitation *P*, ft.	0.22	0.21	0.24	0.25	0.26	0.23	0.24	0.24	0.26	0.25	0.25	0.24	2.90
Turf evapotranspiration *Et*, ft.	0.00	0.00	0.04	0.13	0.28	0.43	0.52	0.46	0.30	0.16	0.06	0.00	2.37
Lake evaporation *El*, ft.	0.12	0.11	0.16	0.28	0.40	0.51	0.55	0.44	0.32	0.21	0.13	0.11	3.35
New York, N.Y.													
Precipitation *P*, ft.	0.30	0.26	0.36	0.31	0.31	0.28	0.31	0.39	0.33	0.29	0.33	0.31	3.79
Turf evapotranspiration *Et*, ft.	0.02	0.02	0.07	0.17	0.33	0.47	0.57	0.50	0.34	0.20	0.08	0.03	2.81
Lake evaporation *El*, ft.	0.15	0.16	0.25	0.36	0.43	0.52	0.55	0.47	0.36	0.27	0.19	0.15	3.86
Syracuse, N.Y.													
Precipitation *P*, ft.	0.17	0.18	0.24	0.23	0.26	0.26	0.27	0.25	0.22	0.24	0.21	0.18	2.72
Turf evapotranspiration *Et*, ft.	0.00	0.00	0.04	0.14	0.29	0.44	0.52	0.46	0.29	0.16	0.06	0.00	2.40
Lake evaporation *El*, ft.	0.09	0.11	0.16	0.29	0.41	0.49	0.54	0.44	0.28	0.19	0.11	0.08	3.18

	Jan	Feb	Mar	Apr	May	Jun	Jul	Aug	Sep	Oct	Nov	Dec	Annual
NORTH CAROLINA													
Raleigh, N.C.													
Precipitation P, ft.	0.31	0.30	0.35	0.31	0.31	0.32	0.47	0.42	0.33	0.25	0.25	0.29	3.91
Turf evapotranspiration Et, ft.	0.05	0.06	0.14	0.27	0.43	0.57	0.63	0.56	0.40	0.24	0.11	0.05	3.53
Lake evaporation El, ft.	0.15	0.17	0.30	0.40	0.44	0.48	0.48	0.42	0.33	0.25	0.18	0.15	3.75
NORTH DAKOTA													
Fargo, N.D.													
Precipitation P, ft.	0.04	0.04	0.06	0.11	0.20	0.29	0.23	0.22	0.14	0.10	0.06	0.04	1.54
Turf evapotranspiration Et, ft.	0.00	0.00	0.00	0.11	0.28	0.42	0.54	0.45	0.25	0.11	0.00	0.00	2.16
Lake evaporation El, ft.	0.08	0.08	0.14	0.29	0.47	0.51	0.59	0.51	0.30	0.24	0.11	0.08	3.38
OHIO													
Cleveland, Ohio													
Precipitation P, ft.	0.23	0.19	0.26	0.29	0.31	0.31	0.30	0.19	0.25	0.23	0.22	0.20	2.98
Turf evapotranspiration Et, ft.	0.00	0.00	0.06	0.16	0.31	0.47	0.53	0.47	0.32	0.17	0.06	0.02	2.56
Lake evaporation El, ft.	0.11	0.10	0.16	0.30	0.41	0.50	0.53	0.44	0.32	0.24	0.14	3.97	7.23
Columbus, Ohio													
Precipitation P, ft.	0.26	0.20	0.28	0.30	0.32	0.35	0.33	0.26	0.22	0.18	0.21	0.21	3.11
Turf evapotranspiration Et, ft.	0.02	0.02	0.08	0.18	0.35	0.50	0.57	0.50	0.33	0.18	0.06	0.02	2.84
Lake evaporation El, ft.	0.11	0.11	0.19	0.31	0.40	0.49	0.51	0.43	0.32	0.22	0.13	0.09	3.32
OKLAHOMA													
Oklahoma City, Ok.													
Precipitation P, ft.	0.12	0.13	0.17	0.29	0.45	0.37	0.26	0.22	0.28	0.24	0.15	0.13	2.82
Turf evapotranspiration Et, ft.	0.04	0.06	0.14	0.28	0.45	0.61	0.71	0.66	0.44	0.26	0.11	0.05	3.82
Lake evaporation El, ft.	0.18	0.21	0.38	0.50	0.53	0.65	0.72	0.69	0.48	0.39	0.23	0.18	5.17
OREGON													
Portland, Ore.													
Precipitation P, ft.	0.64	0.51	0.49	0.27	0.23	0.18	0.04	0.06	0.16	0.40	0.60	0.72	4.32
Turf evapotranspiration Et, ft.	0.03	0.05	0.11	0.18	0.28	0.35	0.44	0.39	0.28	0.16	0.07	0.05	2.39
Lake evaporation El, ft.	0.08	0.10	0.17	0.24	0.33	0.38	0.51	0.41	0.28	0.13	0.08	0.06	2.76
PENNSYLVANIA													
Philadelphia, Pa.													
Precipitation P, ft.	0.27	0.22	0.32	0.29	0.33	0.32	0.37	0.39	0.29	0.26	0.29	0.26	3.63
Turf evapotranspiration Et, ft.	0.02	0.03	0.08	0.19	0.36	0.50	0.59	0.51	0.34	0.19	0.08	0.03	2.94
Lake evaporation El, ft.	0.13	0.15	0.25	0.37	0.43	0.53	0.54	0.46	0.34	0.25	0.17	0.13	3.74
RHODE ISLAND													
Providence, R.I.													
Precipitation P, ft.	0.33	0.27	0.35	0.32	0.27	0.25	0.24	0.34	0.29	0.27	0.36	0.31	3.61
Turf evapotranspiration Et, ft.	0.00	0.02	0.06	0.15	0.29	0.41	0.52	0.46	0.31	0.17	0.07	0.02	2.49
Lake evaporation El, ft.	0.15	0.16	0.23	0.34	0.44	0.49	0.50	0.45	0.32	0.25	0.18	0.12	3.62
SOUTH CAROLINA													
Charleston, S.C.													
Precipitation P, ft.	0.21	0.28	0.32	0.25	0.30	0.39	0.54	0.50	0.44	0.23	0.18	0.23	3.88
Turf evapotranspiration Et, ft.	0.08	0.10	0.20	0.33	0.50	0.61	0.66	0.61	0.46	0.30	0.15	0.09	4.08
Lake evaporation El, ft.	0.18	0.22	0.33	0.43	0.51	0.51	0.51	0.45	0.36	0.29	0.19	0.17	4.15

	Jan	Feb	Mar	Apr	May	Jun	Jul	Aug	Sep	Oct	Nov	Dec	Annual
SOUTH DAKOTA													
Rapid City, S.D.													
Precipitation *P*, ft.	0.03	0.05	0.09	0.14	0.21	0.28	0.14	0.17	0.11	0.08	0.04	0.03	1.37
Turf evapotranspiration *Et*, ft.	0.00	0.00	0.03	0.15	0.31	0.46	0.61	0.53	0.31	0.15	0.03	0.00	2.58
Lake evaporation *El*, ft.	0.11	0.12	0.20	0.36	0.46	0.52	0.67	0.64	0.45	0.32	0.16	0.12	4.14
TENNESSEE													
Memphis, Tenn.													
Precipitation *P*, ft.	0.50	0.39	0.44	0.37	0.35	0.33	0.34	0.25	0.27	0.24	0.36	0.37	4.20
Turf evapotranspiration *Et*, ft.	0.04	0.06	0.14	0.28	0.45	0.59	0.66	0.61	0.42	0.25	0.10	0.05	3.65
Lake evaporation *El*, ft.	0.16	0.19	0.31	0.42	0.52	0.60	0.59	0.56	0.42	0.31	0.20	0.15	4.43
Nashville, Tenn.													
Precipitation *P*, ft.	0.50	0.43	0.45	0.35	0.33	0.31	0.35	0.29	0.26	0.22	0.33	0.39	4.22
Turf evapotranspiration *Et*, ft.	0.04	0.06	0.13	0.26	0.43	0.57	0.64	0.58	0.41	0.24	0.10	0.05	3.51
Lake evaporation *El*, ft.	0.12	0.15	0.26	0.39	0.46	0.54	0.53	0.49	0.37	0.27	0.17	0.12	3.86
TEXAS													
Amarillo, Tex.													
Precipitation *P*, ft.	0.05	0.05	0.06	0.11	0.26	0.20	0.21	0.17	0.16	0.16	0.05	0.06	1.53
Turf evapotranspiration *Et*, ft.	0.04	0.06	0.13	0.25	0.41	0.59	0.65	0.59	0.40	0.24	0.09	0.05	3.50
Lake evaporation *El*, ft.	0.27	0.28	0.50	0.66	0.74	0.77	0.83	0.75	0.54	0.49	0.30	0.24	6.36
Dallas, Tex.													
Precipitation *P*, ft.	0.18	0.20	0.18	0.31	0.40	0.26	0.18	0.16	0.23	0.24	0.18	0.19	2.71
Turf evapotranspiration *Et*, ft.	0.06	0.09	0.19	0.33	0.50	0.65	0.73	0.69	0.49	0.31	0.14	0.08	4.24
Lake evaporation *El*, ft.	0.23	0.28	0.41	0.51	0.58	0.67	0.75	0.70	0.52	0.40	0.25	0.21	5.51
El Paso, Tex.													
Precipitation *P*, ft.	0.05	0.03	0.04	0.05	0.10	0.11	0.15	0.13	0.13	0.10	0.03	0.04	0.97
Turf evapotranspiration *Et*, ft.	0.06	0.09	0.19	0.32	0.48	0.62	0.64	0.59	0.43	0.28	0.13	0.08	3.91
Lake evaporation *El*, ft.	0.28	0.37	0.60	0.79	0.93	0.97	0.82	0.76	0.58	0.50	0.32	0.24	7.17
Houston, Tex.													
Precipitation *P*, ft.	0.28	0.30	0.22	0.28	0.35	0.30	0.36	0.38	0.37	0.30	0.29	0.32	3.74
Turf evapotranspiration *Et*, ft.	0.10	0.13	0.24	0.37	0.54	0.66	0.69	0.65	0.50	0.36	0.18	0.12	4.55
Lake evaporation *El*, ft.	0.17	0.21	0.30	0.37	0.46	0.52	0.55	0.47	0.38	0.31	0.22	0.19	4.16
San Antonio, Tex.													
Precipitation *P*, ft.	0.18	0.20	0.16	0.25	0.31	0.26	0.21	0.21	0.30	0.24	0.18	0.20	2.71
Turf evapotranspiration *Et*, ft.	0.10	0.13	0.24	0.39	0.55	0.66	0.72	0.69	0.52	0.36	0.18	0.12	4.64
Lake evaporation *El*, ft.	0.21	0.24	0.40	0.44	0.51	0.63	0.71	0.65	0.49	0.40	0.26	0.21	5.16
UTAH													
Salt Lake City, Ut.													
Precipitation *P*, ft.	0.13	0.12	0.14	0.15	0.13	0.09	0.05	0.07	0.06	0.11	0.12	0.13	1.31
Turf evapotranspiration *Et*, ft.	0.00	0.02	0.07	0.17	0.30	0.42	0.58	0.50	0.31	0.16	0.05	0.02	2.60
Lake evaporation *El*, ft.	0.08	0.11	0.24	0.39	0.55	0.69	0.87	0.76	0.52	0.32	0.13	0.07	4.71

	Jan	Feb	Mar	Apr	May	Jun	Jul	Aug	Sep	Oct	Nov	Dec	Annual
VERMONT													
Burlington, Vt.													
Precipitation *P*, ft.	0.21	0.18	0.22	0.25	0.29	0.31	0.34	0.28	0.31	0.26	0.26	0.21	3.12
Turf evapotranspiration *Et*, ft.	0.00	0.00	0.03	0.13	0.28	0.42	0.50	0.43	0.26	0.13	0.05	0.00	2.24
Lake evaporation *El*, ft.	0.07	0.08	0.14	0.23	0.36	0.44	0.49	0.39	0.25	0.17	0.09	0.08	2.78
VIRGINIA													
Norfolk, Va.													
Precipitation *P*, ft.	0.28	0.25	0.29	0.19	0.30	0.30	0.45	0.45	0.32	0.25	0.25	0.24	3.57
Turf evapotranspiration *Et*, ft.	0.04	0.05	0.12	0.24	0.41	0.55	0.62	0.56	0.39	0.24	0.11	0.05	3.39
Lake evaporation *El*, ft.	0.18	0.19	0.32	0.42	0.50	0.57	0.55	0.47	0.39	0.26	0.20	0.17	4.22
Richmond, Va.													
Precipitation *P*, ft.	0.29	0.25	0.31	0.29	0.31	0.31	0.42	0.40	0.30	0.24	0.25	0.26	3.63
Turf evapotranspiration *Et*, ft.	0.04	0.05	0.12	0.24	0.40	0.54	0.60	0.54	0.38	0.22	0.10	0.04	3.27
Lake evaporation *El*, ft.	0.12	0.16	0.26	0.39	0.45	0.52	0.52	0.43	0.32	0.22	0.16	0.12	3.67
WASHINGTON													
Seattle, Wash.													
Precipitation *P*, ft.	0.47	0.37	0.33	0.21	0.17	0.16	0.07	0.08	0.17	0.34	0.46	0.53	3.37
Turf evapotranspiration *Et*, ft.	0.03	0.05	0.10	0.18	0.28	0.35	0.42	0.37	0.26	0.15	0.07	0.04	2.30
Lake evaporation *El*, ft.	0.09	0.11	0.18	0.25	0.37	0.41	0.49	0.39	0.26	0.14	0.08	0.08	2.86
Spokane, Wash.													
Precipitation *P*, ft.	0.20	0.15	0.14	0.11	0.14	0.15	0.06	0.07	0.10	0.17	0.19	0.21	1.69
Turf evapotranspiration *Et*, ft.	0.00	0.00	0.06	0.15	0.27	0.36	0.47	0.39	0.24	0.12	0.03	0.00	2.09
Lake evaporation *El*, ft.	0.06	0.08	0.17	0.30	0.40	0.53	0.72	0.58	0.35	0.19	0.06	0.05	3.49
WEST VIRGINIA													
Charleston, W.Va.													
Precipitation *P*, ft.	0.31	0.27	0.35	0.29	0.32	0.34	0.41	0.32	0.24	0.19	0.23	0.25	3.53
Turf evapotranspiration *Et*, ft.	0.03	0.04	0.10	0.22	0.39	0.53	0.59	0.53	0.37	0.21	0.08	0.04	3.11
Lake evaporation *El*, ft.	0.12	0.14	0.23	0.34	0.41	0.48	0.48	0.42	0.32	0.21	0.14	0.11	3.39
WISCONSIN													
Madison, Wisc.													
Precipitation *P*, ft.	0.12	0.10	0.16	0.23	0.28	0.34	0.30	0.30	0.29	0.19	0.18	0.12	2.60
Turf evapotranspiration *Et*, ft.	0.00	0.00	0.04	0.15	0.31	0.47	0.56	0.48	0.30	0.15	0.04	0.00	2.50
Lake evaporation *El*, ft.	0.09	0.10	0.15	0.28	0.39	0.50	0.51	0.43	0.27	0.20	0.09	0.08	3.08
WYOMING													
Cheyenne, Wyo.													
Precipitation *P*, ft.	0.04	0.05	0.08	0.13	0.17	0.15	0.12	0.09	0.08	0.07	0.05	0.04	1.08
Turf evapotranspiration *Et*, ft.	0.00	0.00	0.03	0.11	0.22	0.35	0.47	0.40	0.23	0.12	0.03	0.00	1.96
Lake evaporation *El*, ft.	0.22	0.21	0.27	0.41	0.48	0.57	0.66	0.61	0.46	0.38	0.25	0.22	4.74

Precipitation and Bermuda turf evapotranspiration are from Toro Company, 1966, *Rainfall-Evapotranspiration Data*, Minneapolis: Toro Co.
Lake evaporation is estimated from Kohler method, of W.W. Lamoreaux, 1962, Modern Evaporation Formulae Adapted to Computer Use, *Monthly Weather Review* vol. 90, pages 26-28.

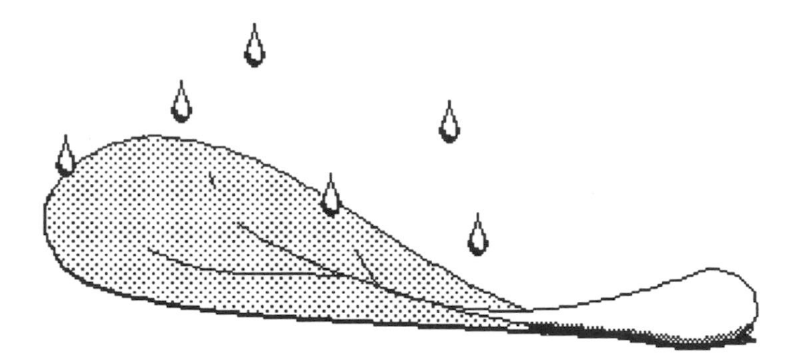

Depth of monthly runoff can be estimated with a rainfall-runoff model:

$$Dr = P\,E$$

where,

Dr = depth of runoff, ft./mo.;
P = Precipitation, ft./mo.; and
E = Efficiency of runoff (ratio of runoff to rainfall, no units): assumed for the present to equal the ratio of SCS runoff depth to rainfall depth in the 2-year storm.

Volume of monthly runoff is simply depth times area:

$$Ro = Dr\,Ac$$

where,

Ro = Runoff, ac.ft./mo; and
Ac = Catchment area, ac.

For precipitation P, long-term averages at U.S. cities are listed in the preceding table. Among the many other sources of local data, Ruffner and Bair's *Weather Almanac*, listed in the *Bibliography*, is convenient for designers working in a variety of regions. Select long-term average values, or values at a recurrence interval apropriate for your application.

Efficiency *E* depends on the watershed's soil and land use. In urban catchments a high runoff efficiency is presumably associated with a high proportion of impervious cover. Reliable ways to estimate monthly runoff efficiency of urban catchments have not yet been developed. In the absence of more direct data the ratio of SCS runoff depth *Dr* to rainfall depth *P* during the 2-year storm can be used as an interim value of runoff efficiency.

Catchment area *Ac* can be scaled from a site map. See the Appendix on *Drainage Area Analysis* if you need help with identifying drainage areas or analyzing their characteristics.

The runoff model on this page was suggested in Dwayne H. Fink and William J. Ehrler, 1984, The Runoff Farming Agronomic System: Applications and Design Concepts, *Hydrology and Water Resources in Arizona and the Southwest* vol. 14, pages 33-40.

Some applications require average monthly runoff in units of ac.ft./day rather that ac.ft./mo. Convert monthly runoff to average daily runoff, so you can work in common units. The conversion is,

$$Ro/\text{day} = (Ro/\text{month})/(\text{Days/month})$$

For number of days per month some workers just use the average of 30 in such conversions. However, the number varies by about 10 percent, from 28.25 (the average for February including leap years) to 31. Accuracy can be improved by using the correct number for each month.

Month	*Days per month*
January	31
February (avg.)	28.25
March	31
April	30
May	31
June	30
July	31
August	31
September	30
October	31
November	30
December	31
Average	30.4

Water Balance Estimation:
1. Runoff, not including snow storage

Site 1

Runoff efficiency:

2-year, 24 hour *P*, in. (from page 35) = ____________ in.
Hydrologic soil group (from page 37) = ____________
Curve Number (from page 39 or 40) = ____________
Runoff depth *Dr* (from page 42) = ____________ in.
Runoff efficiency *E* = *Dr*/*P* = ____________

	Jan	Feb	Mar	Apr	May	Jun	Jul	Aug	Sep	Oct	Nov	Dec	Annual
Monthly runoff depth:													
Precipitation *P*, ft./mo. (from page 58-65)	____	____	____	____	____	____	____	____	____	____	____	____	______
Runoff depth *Dr*, ft./mo. = *P E*	____	____	____	____	____	____	____	____	____	____	____	____	______
Monthly runoff volume:													
Drainage area *Ac*, ac. (from site map)	____	____	____	____	____	____	____	____	____	____	____	____	______
Runoff volume *Ro*, ac.ft./mo. = *Dr Ac*	____	____	____	____	____	____	____	____	____	____	____	____	______
Daily runoff volume:													
Days per month	31	28	31	30	31	30	31	31	30	31	30	31	______
Runoff volume *Ro*, ac.ft./day = Monthly *Ro* / Days	____	____	____	____	____	____	____	____	____	____	____	____	______

Water Balance Estimation:
1. Runoff, not including snow storage

Site 2

Runoff efficiency:

2-year, 24 hour P, in. (from page 35)	=	____ in.
Hydrologic soil group (from page 37)	=	____
Curve Number (from page 39 or 40)	=	____
Runoff depth Dr (from page 42)	=	____ in.
Runoff efficiency $E = Dr/P$	=	____

	Jan	Feb	Mar	Apr	May	Jun	Jul	Aug	Sep	Oct	Nov	Dec	Annual
Monthly runoff depth:													
Precipitation P, ft./mo. (from page 58-65)	___	___	___	___	___	___	___	___	___	___	___	___	___
Runoff depth Dr, ft./mo. $= P E$	___	___	___	___	___	___	___	___	___	___	___	___	___
Monthly runoff volume:													
Drainage area Ac, ac. (from site map)	___	___	___	___	___	___	___	___	___	___	___	___	___
Runoff volume Ro, ac.ft./mo. $= Dr\ Ac$	___	___	___	___	___	___	___	___	___	___	___	___	___
Daily runoff volume:													
Days per month	31	28	31	30	31	30	31	31	30	31	30	31	___
Runoff volume Ro, ac.ft./day = Monthly Ro / Days	___	___	___	___	___	___	___	___	___	___	___	___	___

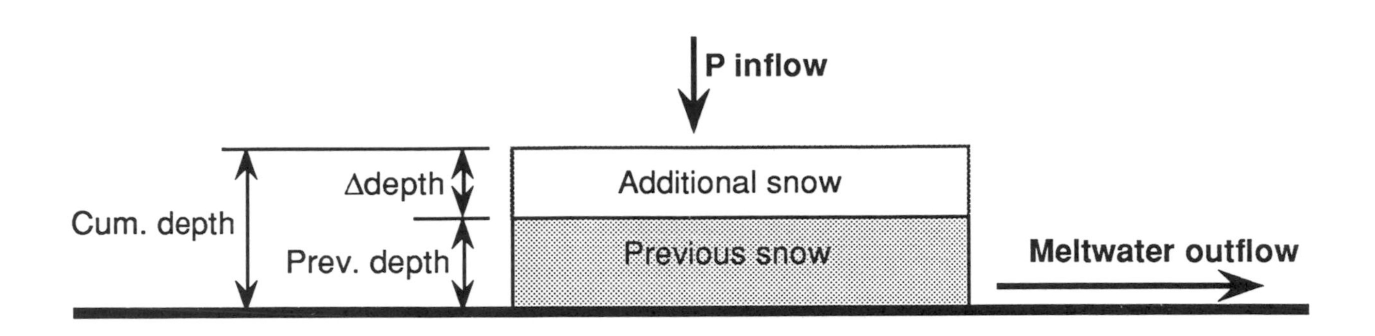

In cold regions a significant proportion of winter precipitation is stored on the surface of watersheds as snow. It should not be counted in runoff estimates until it melts and becomes available to runoff as liquid water.

The effect on runoff can be evaluated with a snow-storage routing procedure. Accumulating snow can be analyzed as a reservoir using the basic reservoir equation, ΔStorage = Inflow - Outflow. Inflow to the snow reservoir is freezing precipitation; outflow is melting snow. It is assumed that for each degree Fahrenheit the monthly temperature falls below 32°F, 10 percent of the monthly precipitation is stored as snow. In warm months, it is assumed that for each degree the temperature rises above 32°F, 10 percent of the accumulated snow melts.

The difference between addition and melt gives each month's net change in snow depth (Δdepth). A positive change means that monthly precipitation is being put into snow storage and not contributing to runoff; a negative change means that snow is melting and adding to monthly runoff. When Δdepth is subtracted from each month's "new" precipitation, the difference is considered the net precipitation available for runoff, and used in place of P in the runoff depth equation, $Dr = PE$. The cumulative snow depth is carried from each cold month to the next, until the entire reservoir melts.

Water Balance Estimation:
1. Runoff, including snow storage

Site 1

Runoff efficiency:
2-year, 24 hour P, in. (from page 35) = ____________ in.
Hydrologic soil group (from page 37) = ____________
Curve Number (from page 39 or 40) = ____________
Runoff depth Dr (from page 42) = ____________ in.
Runoff efficiency $E = Dr/P$ = ____________

	Jan	Feb	Mar	Apr	May	Jun	Jul	Aug	Sep	Oct	Nov	Dec	Annual
Precipitation:													
Precipitation P, ft./mo. (from pages 58-65)	____	____	____	____	____	____	____	____	____	____	____	____	____
Snow storage:													
Cum. snow depth (ft., ≥ 0) = prev.depth + Δdepth	____	____	____	____	____	____	____	____	____	____	____	____	
Proportion S of P to snow storage (function of temperature)	____	____	____	____	____	____	____	____	____	____	____	____	
Addition to storage(ft.) $= S\,P$	____	____	____	____	____	____	____	____	____	____	____	____	____
Proportion M of snow melting (function of temperature)	____	____	____	____	____	____	____	____	____	____	____	____	
Snow melt (ft.) $= M$ (cum.depth)	____	____	____	____	____	____	____	____	____	____	____	____	____
Δdepth of storage (ft.) = addition - melt	____	____	____	____	____	____	____	____	____	____	____	____	____
Runoff depth:													
Net P avail. for runoff (ft.) $= P - \Delta$ snow depth	____	____	____	____	____	____	____	____	____	____	____	____	____
Runoff depth Dr, ft./mo. = (Net P) E	____	____	____	____	____	____	____	____	____	____	____	____	____
Monthly runoff volume:													
Drainage area Ac, ac. (from site map)	____	____	____	____	____	____	____	____	____	____	____	____	____
Runoff volume Ro, ac.ft./mo. $= Dr\,Ac$	____	____	____	____	____	____	____	____	____	____	____	____	____
Daily runoff volume:													
Days per month	31	28	31	30	31	30	31	31	30	31	30	31	____
Runoff volume Ro, ac.ft./day = Monthly Ro / Days	____	____	____	____	____	____	____	____	____	____	____	____	____

Water Balance Estimation:

1. Runoff, including snow storage

Site 2

Runoff efficiency:

2-year, 24 hour P, in. (from page 35) = ________ in.
Hydrologic soil group (from page 37) = ________
Curve Number (from page 39 or 40) = ________
Runoff depth Dr (from page 42) = ________ in.
Runoff efficiency $E = Dr/P$ = ________

	Jan	Feb	Mar	Apr	May	Jun	Jul	Aug	Sep	Oct	Nov	Dec	Annual
Precipitation:													
Precipitation P, ft./mo. (from pages 58-65)	___	___	___	___	___	___	___	___	___	___	___	___	___
Snow storage:													
Cum. snow depth (ft., ≥ 0) = prev.depth + Δdepth	___	___	___	___	___	___	___	___	___	___	___	___	
Proportion S of P to snow storage (function of temperature)	___	___	___	___	___	___	___	___	___	___	___	___	
Addition to storage(ft.) = $S\ P$	___	___	___	___	___	___	___	___	___	___	___	___	___
Proportion M of snow melting (function of temperature)	___	___	___	___	___	___	___	___	___	___	___	___	
Snow melt (ft.) = M (cum.depth)	___	___	___	___	___	___	___	___	___	___	___	___	___
Δdepth of storage (ft.) = addition - melt	___	___	___	___	___	___	___	___	___	___	___	___	___
Runoff depth:													
Net P avail. for runoff (ft.) = P - Δ snow depth	___	___	___	___	___	___	___	___	___	___	___	___	___
Runoff depth Dr, ft./mo. = (Net P) E	___	___	___	___	___	___	___	___	___	___	___	___	___
Monthly runoff volume:													
Drainage area Ac, ac. (from site map)	___	___	___	___	___	___	___	___	___	___	___	___	___
Runoff volume Ro, ac.ft./mo. = $Dr\ Ac$	___	___	___	___	___	___	___	___	___	___	___	___	___
Daily runoff volume:													
Days per month	31	28	31	30	31	30	31	31	30	31	30	31	___
Runoff volume Ro, ac.ft./day = Monthly Ro / Days	___	___	___	___	___	___	___	___	___	___	___	___	___

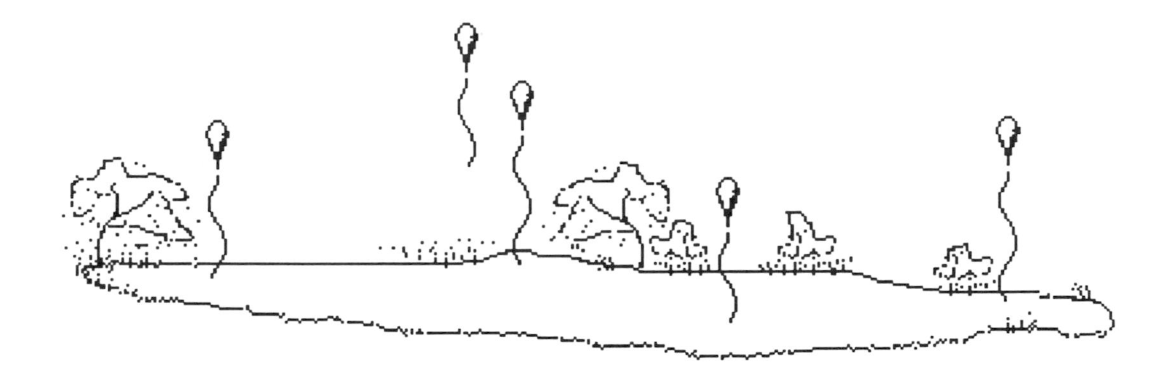

Evaporation and evapotranspiration are measured in units of depth per month. The depth does not vary with size of watershed.

Evaporation from an open water surface is symbolized *El* to designate "lake" evaporation. It applies when you are working with a body of standing water such as the permanent pool of a "wet" detention basin. Monthly estimates by the authors for U.S. cities are listed in the table on pages 58-65. Those estimates are based on the Kohler method, which is a very laborious method although it is probably the simplest of the acceptably accurate estimation methods. U.S. maps by Adolph Meyer, also based on the Kohler method, were published in the 1961 edition of Chapter 9, "Storage Requirements for Beneficial Use," of SCS's *National Engineering Handbook, Section 4, Hydrology*. For information about the Kohler method, see the references by Kohler and others or Dunne and Leopold, listed in the *Bibliography*. Another kind of estimation method is to apply a coefficient to evaporation from a Class A evaporation pan, data for which are widely available. The most common value of the coefficient is 0.7, but it can vary from place to place and month to month.

Evapotranspiration is symbolized *Et*. It applies to a water balance when you are working with a vegetated soil area such as a "dry" stormwater basin. Monthly estimates by the Toro Company for U.S. cities are listed in the table on pages 58-65. Those estimates are based on the "Blaney-Criddle" method for Bermuda grass. The Blaney-Criddle method is a simple method which has the advantage of taking different types of proposed plantings into account. For information about the Blaney-Criddle method, see Ferguson's chapter on "Effective Use of Water" in *Irrigation*, volume 3 of the Landscape Architecture Foundation's *Landscape Architectural Construction Handbook*. Another kind of estimate is "potential evapotranspiration" such as the Thornthwaite method described in Dunne and Leopold's book; *PEt* does not take into account specific types of plantings.

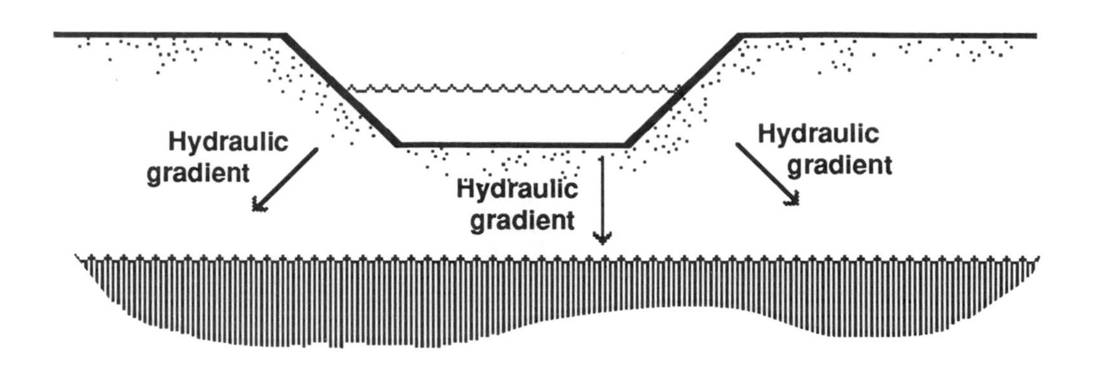

Infiltration is governed by Darcy's equation. Darcy was a Frenchman who discovered this relationship about a century and a half ago. Since then this relationship has been richly confirmed, and found to have almost universal application to subsurface flow. It is a simple relationship:

$$Q = A\,K\,Gh$$

where,

Q = infiltration, ac.ft./day;
A = cross-sectional area through which the water infiltrates, ac.;
K = saturated hydraulic conductivity, the permeability or infiltration rate of the soil, ft./day (see the table on the next page); and
Gh = hydraulic gradient (no units) = $\Delta H / l$, where,

> ΔH = difference in head (pressure), in feet of water, between two points in the path of the water's movement; and
> l = distance along path of movement, in feet.

Through a level basin floor, the direction of movement of infiltrating water is likely to be close to vertical. Thus l is equal to the distance of water's "fall" into the soil. If the underlying soil is homogeneous with no restraining soil layer or groundwater table, then the loss of head ΔH is is also equal to the water's loss of elevation into the soil. Thus ΔH and l are equal, and Gh is equal to 1.0.

Through a basin's sides, the direction of movement of infiltrating water is unlikely to be vertical if the side slope exceeds about 20 percent (5:1). The movement is more likely to be at a low angle, in which case the hydraulic gradient Gh would be closer to 0.5 than to 1.0.

Saturated hydraulic conductivity ***K*** depends on site-specific soil types. It can be estimated from soil borings or *in situ* tests such as with double-barrel infiltrometers. Such tests involve expenses, and do not always give consitent results. An alternative method is to identify soil types from SCS soil surveys or (preferably) on-site examination, and then to estimate *I* by association with soil texture. The table below lists average conductivities found in the laboratory by highway drainage engineers and USDA soil scientists. When on-site data are extremely limited, only very conservative assumptions would be prudent: for a wet pond, assume soil texture with permeability as high as can be found in the region; for a dry basin, assume texture with permeability as low as can be found.

	Hydraulic conductivity K	
	in/hr	*ft/day*
Crushed stone: ASTM stone size:		
No. 3 stone	50,000	100,000
No. 4 stone	40,000	80,000
No. 5 stone	25,000	50,000
No. 6 stone	15,000	30,000
Natural soil: SCS texture class:		
Sand	8.27	16.54
Loamy sand	2.41	4.82
Sandy loam	1.02	2.04
Loam	0.52	1.04
Silt loam	0.27	0.54
Sandy clay loam	0.17	0.34
Clay loam	0.09	0.18
Silty clay loam	0.06	0.12
Sandy clay	0.05	0.10
Silty clay	0.04	0.08
Clay	0.02	0.04

Stone data inferred from H.R. Cedergren and others, 1972, *Guidelines for the Design of Subsurface Drainage Systems for Highway Structural Sections*, Report No. FHWA-RD-30, Washington: Federal Highway Administration Office of Research and Development; and Table 2.1 of H.R. Cedergren, 1977, *Seepage, Drainage, and Flow Nets*, New York: Wiley.
Soil conductivities from W.J. Rawls, D.L. Brakensiek and K.E. Saxton, 1982, Estimation of Soil Water Properties, *Transactions of the American Society of Agricultural Engineers*, vol. 25, no. 5, pages 1316-1320 and 1328.

Water Balance Estimation:
2. Evapotranspiration and infiltration, after development

Site 1

	Jan	Feb	Mar	Apr	May	Jun	Jul	Aug	Sep	Oct	Nov	Dec	Annual
Evapotranspiration:													
Turf evapotrans. *Et*, ft./mo. (from page 58-65)													
Lake evaporation *El*, ft./mo. (from page 58-65)													
Monthly infiltration:													
Days per month	31	28	31	30	31	30	31	31	30	31	30	31	365
Soil infiltration rate *K*, ft./day (from page 75)													
Monthly infiltration *I*, ft./mo. = *K* (days/mo.)													

Site 2

	Jan	Feb	Mar	Apr	May	Jun	Jul	Aug	Sep	Oct	Nov	Dec	Annual
Evapotranspiration:													
Turf evapotrans. *Et*, ft./mo. (from page 58-65)													
Lake evaporation *El*, ft./mo. (from page 58-65)													
Monthly infiltration:													
Days per month	31	28	31	30	31	30	31	31	30	31	30	31	365
Soil infiltration rate *K*, ft./day (from page 75)													
Monthly infiltration *I*, ft./mo. = *K* (days/mo.)													

Water balance exercise:
Discussion of results

1. Point out similarities between the monthly runoff equation $Ro = PAcE$ and the "rational" equation for storm runoff. Point out differences.

2. Which month has the maximum monthly runoff on each site? Which month has the minimum? What regional climatic conditions contribute to this seasonal pattern of runoff?

3. Which month has the maximum monthly evaporation and/or evapotranspiration on each site? Which month has the minimum? What regional climatic conditions contribute to this seasonal pattern of evaporation and evapotranspiration?

4. Which month has the greatest water *surplus*, the excess of precipitation over evaporation? Which month has the greatest water *deficit*, the excess of evaporation over precipitation? How do vegetation, streamflow, lake levels and other environmental factors in your area respond to this seasonal pattern?

5. If your site is in a cold region, estimate runoff twice, once using snow storage and once disregarding it. What change in magnitude of *maximum* monthly runoff results from taking snow storage into account? In what way does the month in which the maximum runoff occurs change? What change in magnitude of *minimum* monthly runoff occurs? In what way does the month in which the minimum runoff occurs change?

6. Using algebra, derive an equation for velocity of infiltrating water from the form of Darcy's equation on page 74. What kinds of design criteria could such a velocity equation help to fulfill in design of a site-specific basin?

Summary and Commentary

Design storms are isolated, rare, brief and extreme.

In contrast, monthly inflows and outflows are continuous. Stormwater basin hydrologic regimes and environmental effects are a result of long-term flows and their fluctuation from season to season, not isolated design storms.

Additional flows not taken into account in this chapter may occur in some basins, such as irrigation withdrawals or throughflows needed to maintain aeration for aquatic life. To take them into account appropriate additional terms would need to be added to monthly water-balance equations.

The best approaches to water-balance estimation that are currently available are presented in this chapter. We hope that future research will improve some of the quantitative water-balance factors. One area needing refinement is the runoff efficiency E, which, for any given catchment, might vary substantially from month to month. Data from small urban watersheds need to be analyzed to generate a reliable table of monthly urban E values.

Chapter 4

Conveyance is design to move water away in order to avoid on-site nuisances.

Linear pipes and swales draining one into another characterize conveyance systems. Pipes are usually more expensive than swales of similar capacity, because materials for pipes must be purchased whereas swales are often just formed out of earth. But pipes are essential where water has to be carried under roads or where site development is too intense to leave room for relatively broad earth swales.

The key to conveyance is its continuity. In plan view each conveyance connects to adjacent ones. In profile view the *invert* elevation — the bottom elevation at which water flows — slopes continuously downward through pipes and swales alike. The movement of water through those structures is equally continuous. By definiton, conveyance does not prevent a given volume of water from moving across the land surface.

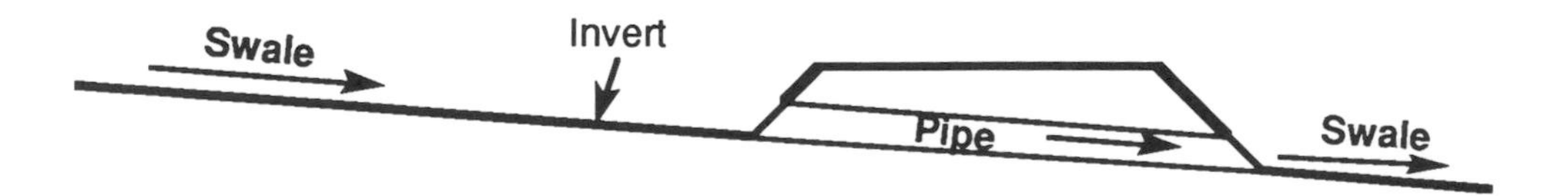

Conveyance is one of the most ancient of stormwater management functions. In the 2,000 year old streets of Pompeii you can see systems of gutters that drain the city's stormwater out to the rivers and the sea.

Today, we could ask whether we really want to throw away fresh water so quickly and irrevocably, and whether the disposed water is hurting anyone downstream. Nevertheless you can see conveyances systems being built in every part of the United States. In parts of many developments conveyance is in fact the only reasonable thing to do and alternative measures are not worth the effort.

A conveyance should be capable of carrying the peak rate of flow from a selected design storm. If it is big enough carry the peak flow, then it can handle all the rest of a design-storm event, and any other lower-level flows, as well. Preventing water from backing up is the whole idea, because backed-up water could overflow and cause a nuisance. Usually you want to make a pipe or swale exactly the size necessary for adequate capacity because larger sizes tend to involve greater construction costs

However, constraints on the sizes or materials of conveyances are sometimes brought about by site-specific factors. When such constraints are found, you have to consider alternative dimensions, materials or alignments of the conveyances, or supplement them with additional hydrologic functions to cut down peak rates of flow.

The depth of water in a swale might be limited by the floor elevations of nearby houses or the pavement elevations of a nearby road. This condition could be recognized by examining a topographic map of the site.

The velocity of flow in a swale, or emerging from a culvert into a swale, might need to be limited in order not to erode the swale's soil. A stable conveyance should not be abraded during the storm for which it is designed. Erosive velocity can be recognized by estimating velocity in a swale during the design storm, and comparing the result with the maximum noneroding velocity for the swale's soil type. Typical maximum noneroding velocities are listed on page 82. From the general velocity equation $V = Q/A$ the velocity of any conveyance can be derived, taking the area A from the size and shape of the cross-section, and flow Q from storm runoff estimates. This chapter includes a specially derived velocity formula for each type of conveyance discussed. An alternative way to estimate velocity is from Manning's equation, which is introduced on page 83 and solutions of which are shown on pages 86 and 87.

To meet depth or velocity constraints you have a number of options. You can find a conveyance with larger capacity, or alter controllable characteristics of the conveyance such as slope. You can increase the stability of a swale by lining it with concrete, masonry, riprap or selected vegetation. Or you can revise your site plan to reduce the Qp that reaches your conveyance.

A storm larger than your design storm can cause a conveyance to overflow. Although you accepted this risk in selecting your design storm, you still have to provide, by appropriate grading, an *emergency overflow* — a nonerodible, nondamaging path for the excess water to follow. Thus there are actually two drainage systems on almost every site: a *primary* system which handles all storms up to the design storm and a *secondary* system for larger storms.

Water quality can be affected by swales to some degree. In the Washington, D.C. area, trace metals were found to have accumulated in grassed swales along highways and residential roads, implying that the metals had been removed to some degree from storm water before it passed into streams or groundwater. The swales had been in place for about 10 years. The percentage removed from the stormwater is not known. The accumulation was greatest closest to the pavement surfaces. All of the accumulation was within the top few inches of the fine-textured soil. The accumulated amount of metals was below any level that might be considered harmful to the soil or to the surrounding environment.

Data about trace-metal accumulation are from Parker J. Wigginton, Clifford W. Randall and Thomas J. Grizzard, 1986, Accumulation of Selected Trace Metals in Soils of Urban Runoff Swale Drains, *Water Resources Bulletin* vol. 22, no. 1, pages 73-79.

Maximum Noneroding Velocities

	On erosion-resistant soil (stiff clay, silt, hardpan, fine & coarse gravel)			On easily eroded soil (fine sand, sandy loam, loam, silt & clay loam)		
	Channel slope			*Channel slope*		
	0-5%	5-10%	10%+	0-5%	5-10%	10%+
Grass cover: Values assume that grass cover is uniform and well maintained.						
Bermudagrass	8 fps	7 fps	6 fps	6 fps	5 fps	4 fps
Buffalograss	7 fps	6 fps	5 fps	5 fps	4 fps	3 fps
Kentucky bluegrass	7 fps	6 fps	5 fps	5 fps	4 fps	3 fps
Smooth brome	7 fps	6 fps	5 fps	5 fps	4 fps	3 fps
Blue grama	7 fps	6 fps	5 fps	5 fps	4 fps	3 fps
Grass mixture	5 fps	4 fps	*erodible*	4 fps	3 fps	*erodible*
Lespedeza sericea	3.5 fps	— *erodible* —		2.5 fps	— *erodible* —	
Weeping lovegrass	3.5 fps	— *erodible* —		2.5 fps	— *erodible* —	
Yellow bluestem	3.5 fps	— *erodible* —		2.5 fps	— *erodible* —	
Kudzu	3.5 fps	— *erodible* —		2.5 fps	— *erodible* —	
Crabgrass	3.5 fps	— *erodible* —		2.5 fps	— *erodible* —	
Common lespedeza	3.5 fps	— *erodible* —		2.5 fps	— *erodible* —	
Sudangrass	3.5 fps	— *erodible* —		2.5 fps	— *erodible* —	
Crushed stone: National Stone Association sizes are based on square openings.						
Min.1", Average 1.5": NSA No. R-2	4.5 fps	4.5 fps	4.5 fps	4.5 fps	4.5 fps	4.5 fps
Min.2", Average 3": NSA No. R-3	6.5 fps	6.5 fps	6.5 fps	6.5 fps	6.5 fps	6.5 fps
Min.3", Average 6": NSA No. R-4	9.0 fps	9.0 fps	9.0 fps	9.0 fps	9.0 fps	9.0 fps

From Tables 3 and 4 of U.S. Federal Highway Administration, 1973, *Design of Roadside Drainage Channels*, and Table 3 of National Crushed Stone Association, 1982, *Quarried Stone for Erosion and Sediment Control*.

Uniform flow exists in a swale or, more rarely, a culvert, where water is delivered at a steady rate, there are no sudden changes in gradient or direction, and the size of pipe or swale is either uniform or increases gradually downstream.

Manning's equation describes uniform flow. Manning was an Irishman who worked about a hundred years ago. In the laboratory you can recreate appropriate conditions and derive essentially the same relationship he did. Despite the age of Manning's equation it is still widely accepted — as long as you apply it only where the flow is free, continuous and uniform.

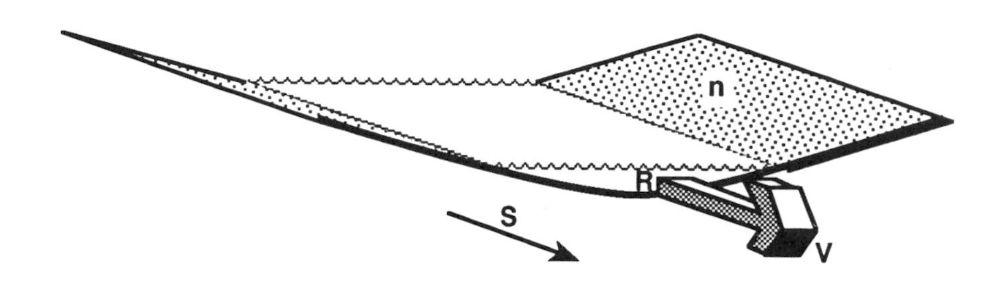

Manning's equation is:

$$V = (1.49/n)\, R^{2/3} S^{1/2}$$

where,

V = velocity (feet per second);
n = roughness factor (no units): see the table on the next page;
R = hydraulic radius = A/Wp, where,
  A = cross-sectional area (sq.ft.);
  Wp = wetted perimeter (ft.), the cross-sectional length of surface in contact with water; and
S = slope along conveyance's length (decimal fraction).

Roughness *n* indicates how much a material resists flow. The values in the table on the next page range from 0.012 to 0.5, so this factor alone could have more than a fortyfold effect on capacity of a conveyance to carry a given flow.

Hydraulic radius *R* indicates how much cross-sectional area is available for carrying water, for a given area of frictional contact with the conveyance's sides. A conveyance that is broad and shallow has a small cross-sectional area in proportion to the wetted perimeter. The larger the radius, the more water can flow through.

Slope *S* indicates how directly gravity can pull water through a conveyance. The greater the slope, the greater the flow.

The charts on pages 86 and 87 illustrate solutions of Manning's equation for V.

The equation can also be solved for *Q*. Since $V = Q/A$ we can substitute for V in the original equation and derive,

$$Q = (1.49/n)\, A\, R^{2/3} S^{1/2}$$

Roughness Factors

Conveyance material	Roughness factor *n* in Manning's equation
Concrete pipe	0.013
Corrugated metal pipe	0.024
Brick	0.014 - 0.017
Concrete swale, trowel finish	0.012 - 0.014
Random stone in mortar	0.020 - 0.023
Dry rubble (riprap)	0.023 - 0.033
1"-1.25" crushed stone on 1-2% slope	0.024
1"-1.25" crushed stone on 10% slope	0.055
2.2"-2.8" crushed stone on 2% slope	0.025
2.2"-2.8" crushed stone on 8-10% slope	0.03
Asphalt	0.013 - 0.016
Earth with short grass, few weeds (turf)	0.022 - 0.027
Earth with dense weeds and high brush	0.08 - 0.12
Earth, clean bottom, brush on sides	0.05 - 0.08
Winding natural streams considerably covered with small growth	0.035
Mountain streams in clean loose cobbles; streams with bank or aquatic vegatation	0.040 - 0.050
Streams with irregular alignment and cross section, obstructed by trees and brush	0.100
Streams with very irregular alignment and cross section, many roots, trees, logs, drift on bottom	0.150-0.200

From Table 2 of U.S. Federal Highway Administration, 1973, *Design of Roadside Drainage Channels*;
Table 16-1 of Dunne and Leopold, 1978, *Water in Environmental Planning*;
Table B-6, page 577, of U.S. Bureau of Reclamation, 1974, *Design of Small Dams*; and
Table 2 of Steven R. Abt and others, 1988, Resistance to Flow over Riprap in Steep Channels, *Water Resources Bulletin* vol. 24, no. 6, pages 1193-1200
(supplemented by personal communication from Steven R. Abt, September 12, 1989).

Manning's equation is complicated enough that practitioners seldom try to solve it by hand. On the following pages are charts based upon Manning's equation, which solve the equation for common applications. A computer program containing Manning's equation would be another way to work out site-specific calculations.

The velocity charts (pages 86 and 87) solve Manning's equation for velocity at four different roughness (n) values, representing four different types of pipe and swale materials.

The pipe capacity charts (pages 88 and 89) are for two common pipe materials, concrete and corrugated metal. Each curve on the charts is for a commercially available diameter of pipe.

The swale capacity charts (pages 90 and 91) are for two common cross-sectional shapes, a triangle and a trapezoid. Each line on the charts is for a different material and depth of flow. If you work with a swale shape not shown on the charts, you can still use one of our charts as long as, for a given roughness and slope, the swale referred to in the chart can fit inside the cross-section of the swale you are working with. If the swale shown on the chart has sufficient capacity to carry the needed flow then the larger swale on your site has more than enough capacity.

On each pipe and swale chart the scale at the bottom shows slope in percent. The scale at the left shows rate of flow in cfs. On the appropriate chart find the point where your *Qp* meets your gradient. If your *Qp* and gradient meet exactly on a line representing a specific size of conveyance, that can be your conveyance. If they do not meet exactly on a line, move up and to the left to find a conveyance with greater capacity. For pipes this amounts to choosing a larger diameter. For swales you have a choice of material as well as of depth, so you can choose one that minimizes cost or fits with the character of your site design.

If you cannot find an acceptable combination of capacity and cost you can modify your site plan to increase the conveyance's gradient or reduce the *Qp* that reaches it.

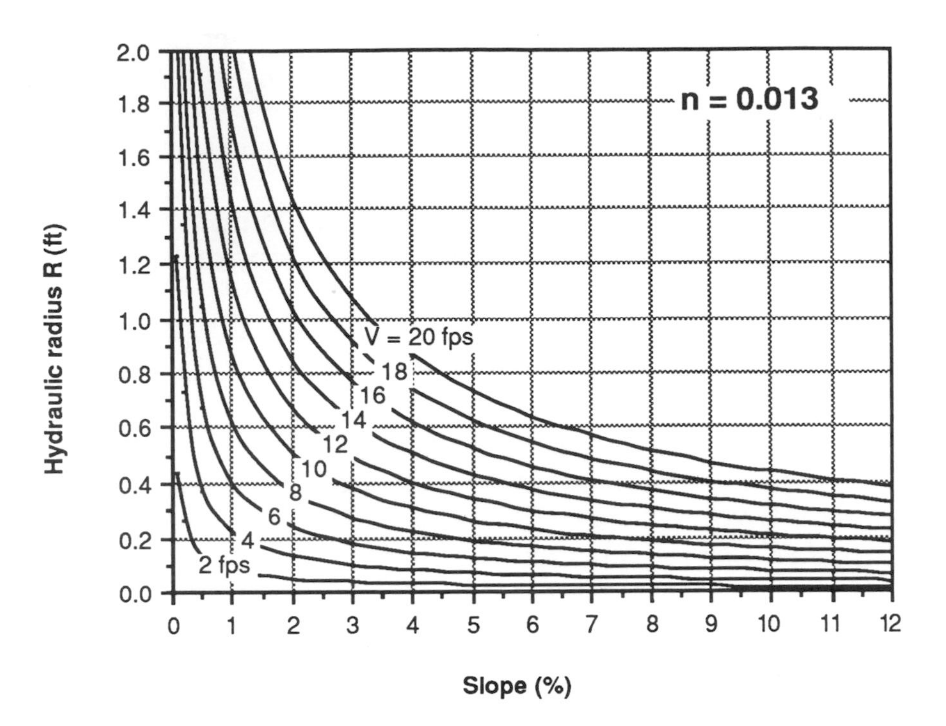

0.013 is a typical n value for concrete. Chart was derived from Manning's equation.

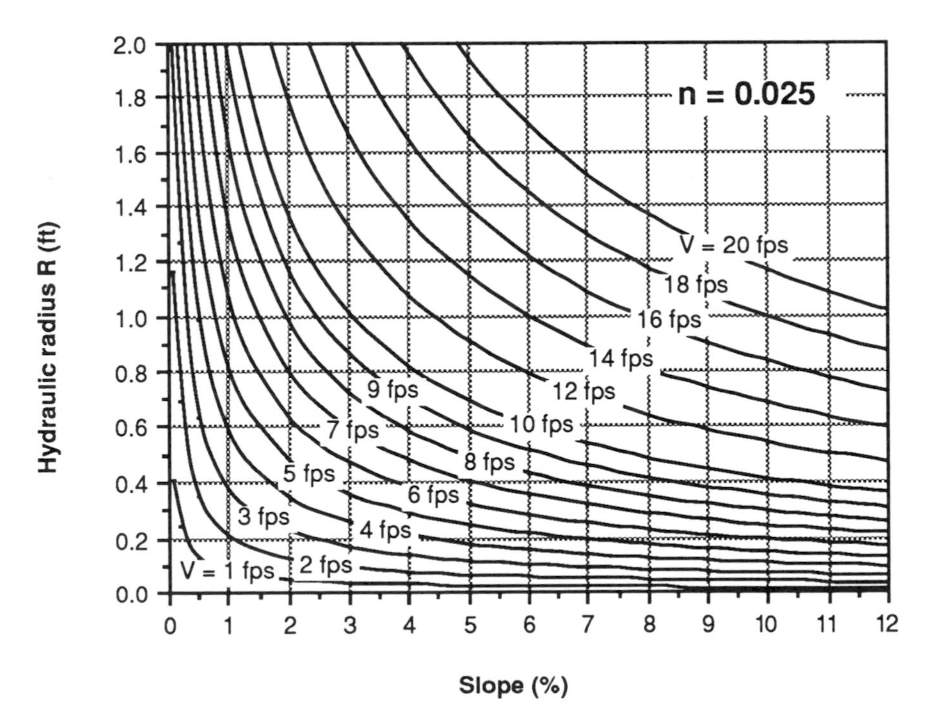

0.025 is a typical n value for turf. Chart was derived from Manning's equation.

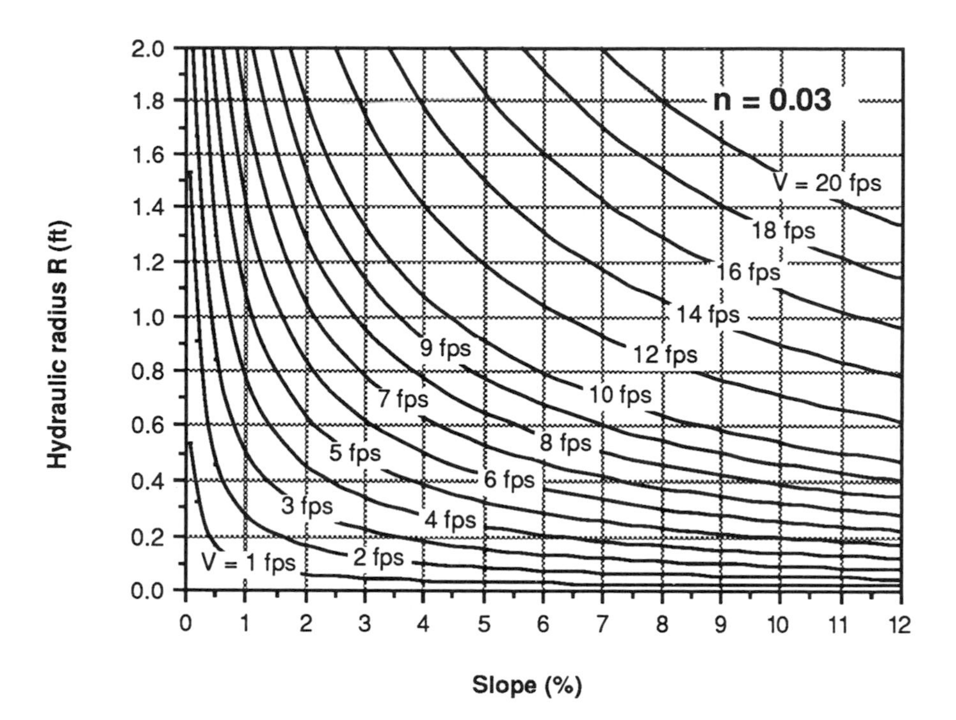

0.03 is a typical n value for riprap. Chart was derived from Manning's equation.

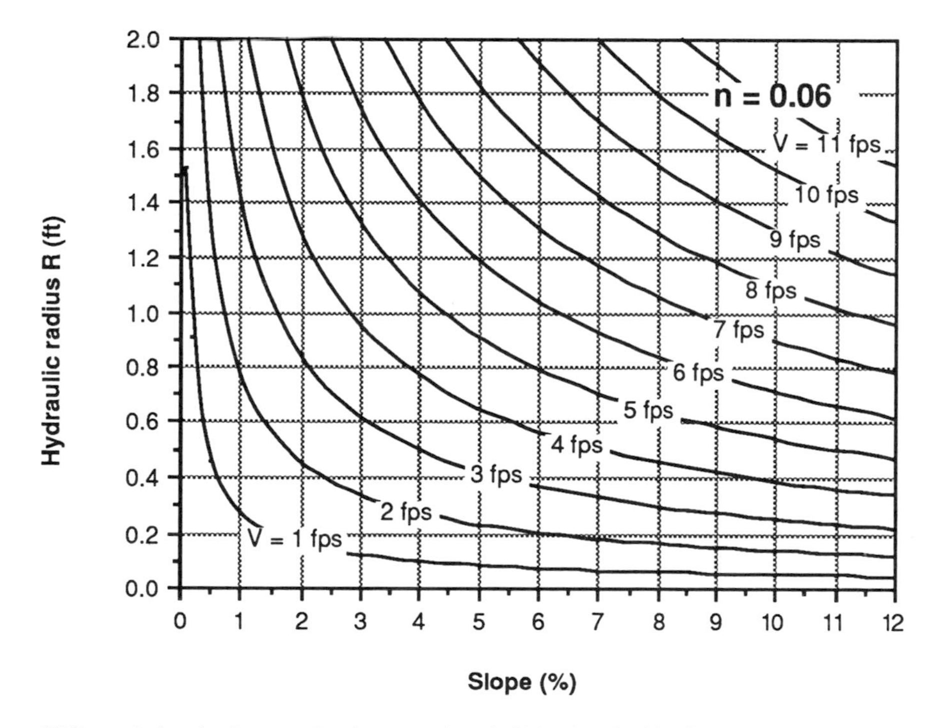

0.06 is a typical n value for a natural earthen stream channel with brush on the sides. Chart was derived from Manning's equation.

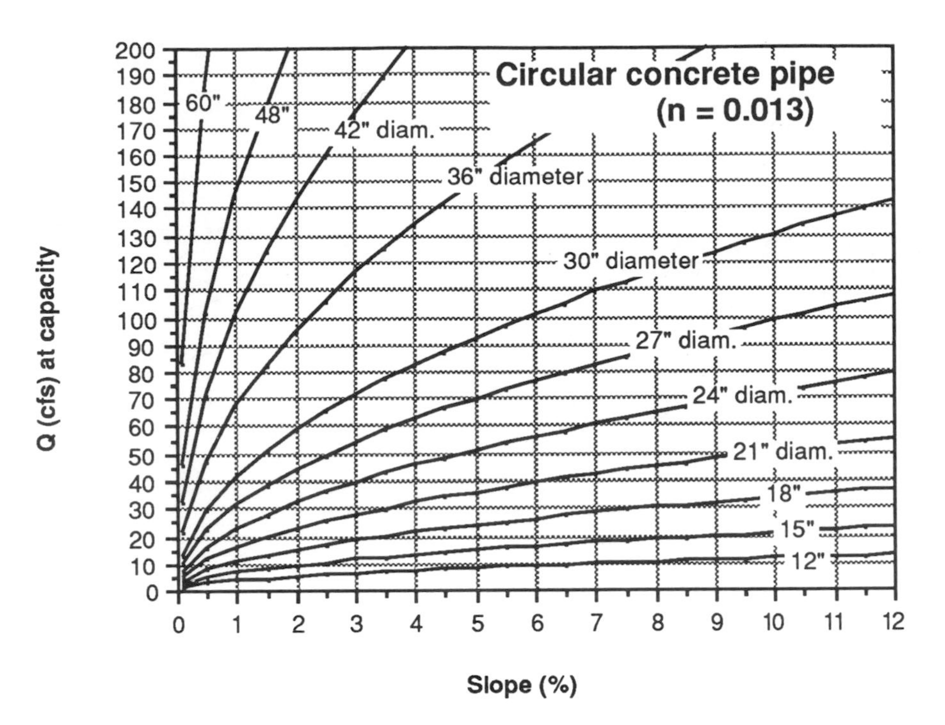

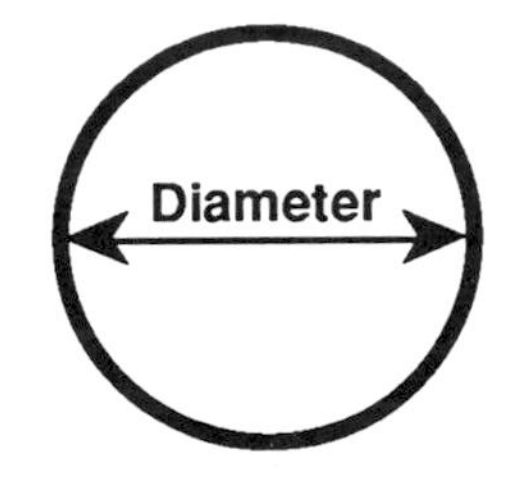

Velocity in a full-flowing circular pipe is given by,

$$V = 183\, Q / D^2$$

where,

V = velocity in feet per second (fps);
Q = rate of flow in cubic feet per second (cfs); and
D = diameter of pipe in inches.

Chart was derived from Manning's equation.
Velocity formula was derived from the equation, $Q = VA$.

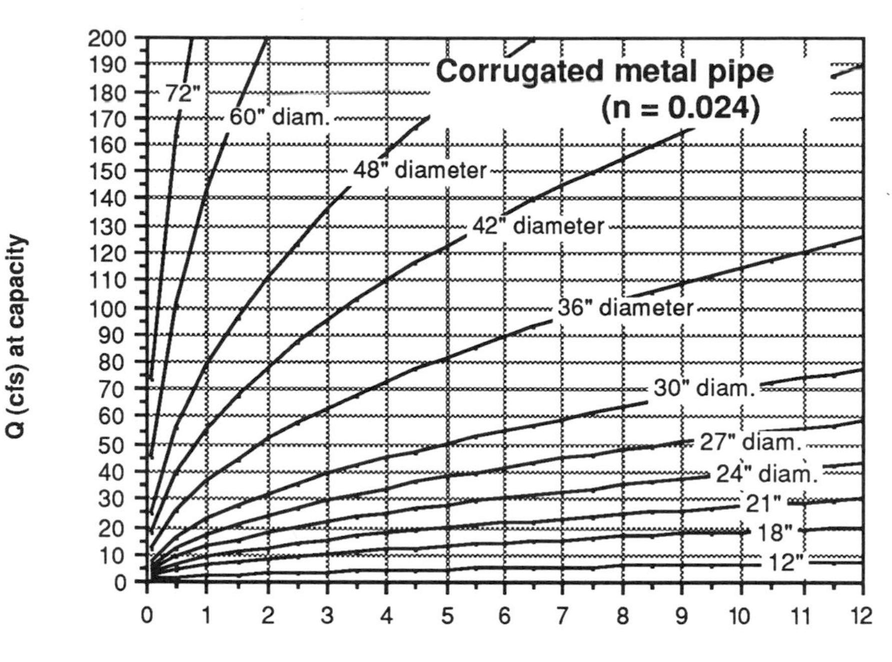

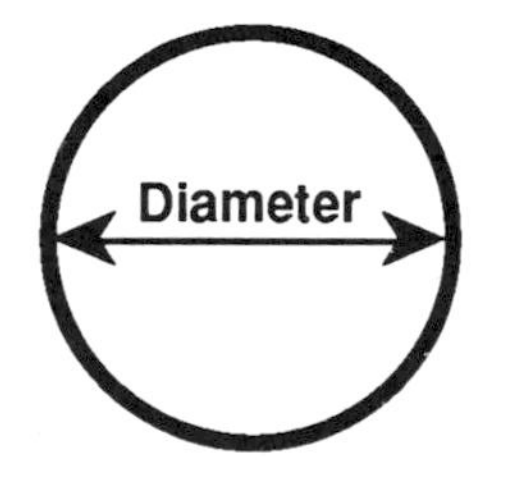

Velocity in a full-flowing circular pipe is given by,

$$V = 183\ Q / D^2$$

where,

V = velocity in feet per second (fps);
Q = rate of flow in cubic feet per second (cfs); and
D = diameter of pipe in inches.

Chart was derived from Manning's equation.
Velocity formula was derived from the equation, $Q = VA$.

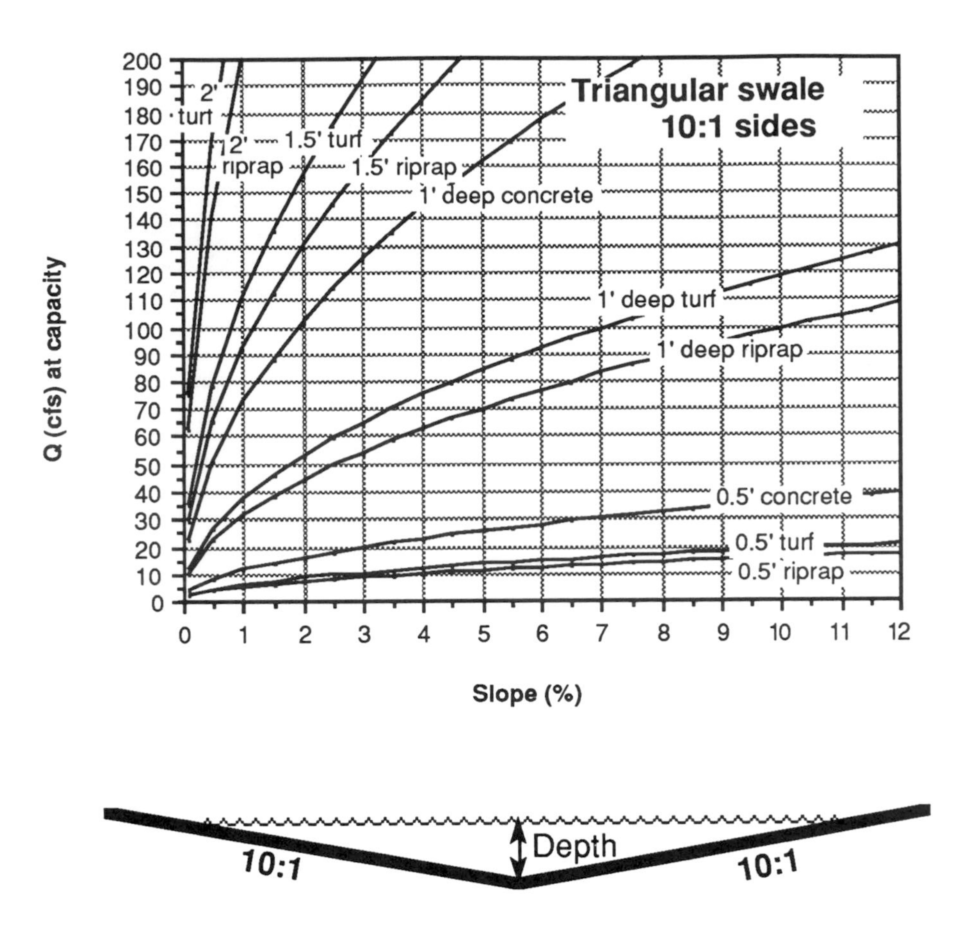

Velocity in a swale of this shape is given by,

$$V = Q / (10\,D^2)$$

where,

V = velocity in fps;
Q = flow in cfs; and
D = depth at deepest (center) point of swale, in ft.

Chart was derived from Manning's equation, with the following roughness (n) values: concrete = 0.013; turf = 0.025; riprap = 0.03.
Velocity formula was derived from equation $Q = V A$.

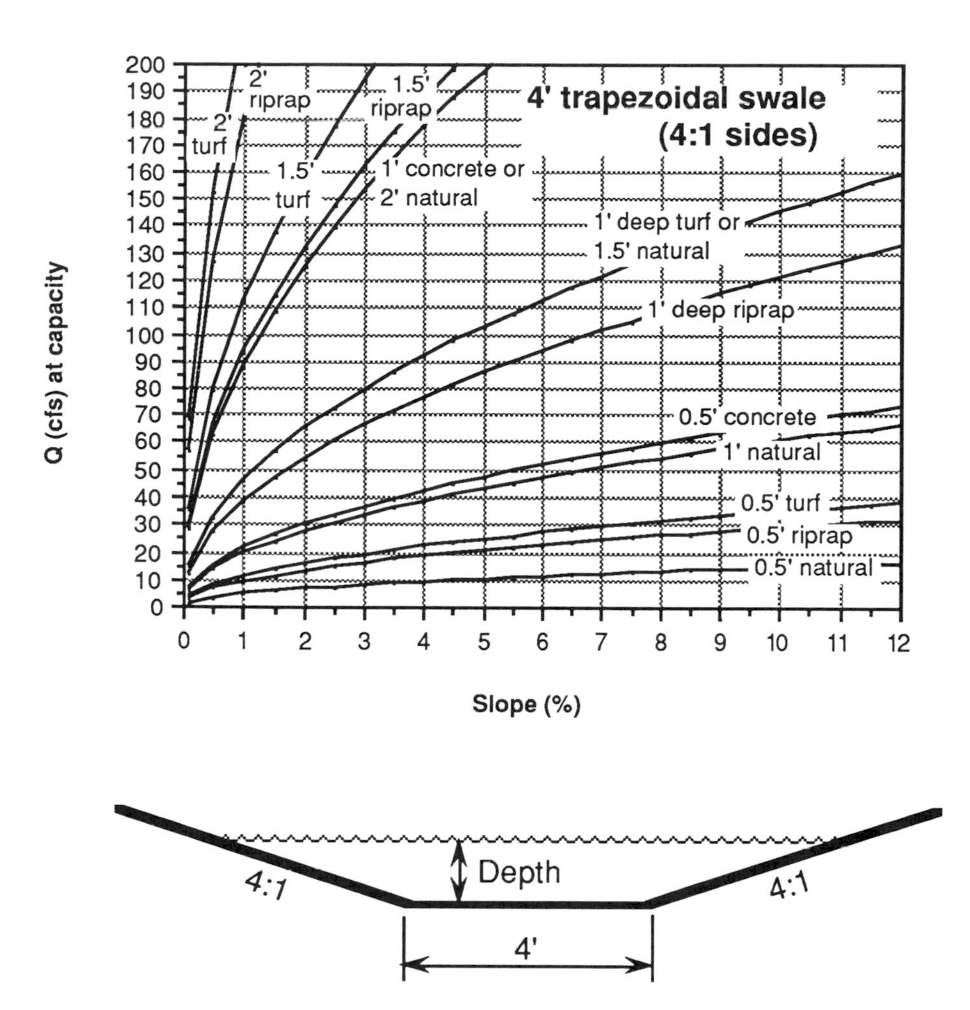

Velocity in a swale of this shape is given by,

$$V = Q / (4D + 4D^2)$$

where,

V = velocity in fps;
Q = rate of flow in cfs; and
D = depth in center of swale, in feet.

Chart was derived from Manning's equation with the following roughness (n) values: concrete = 0.013; turf = 0.025; riprap = 0.03; natural = 0.06.
Velocity formula was derived from the equation, $Q = VA$.

Inlet Control

Inlet control exists where water enters some culverts. The mouth of a pipe is a constriction in water's path, so water might back up behind the pipe before entering. Inlet control applies where water is backed up at the mouth of the culvert but not at the downstream end. The entrance to every culvert should be checked for its inlet-control capacity.

The *deeper* the water the greater the pressure at the bottom. Hydraulic pressure is often referred to as *head*. That strange term seems to have originated in old watermills, where there was *headwater* above the mill and *tailwater* below, and it was convenient to describe available hydraulic pressure in terms of feet of *head* coming down the pipe.

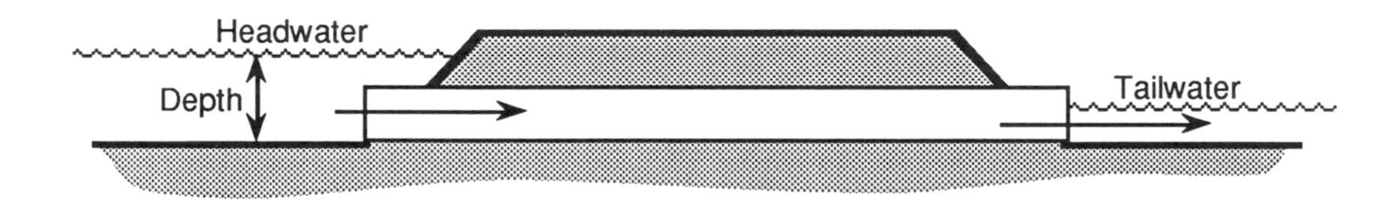

After water has passed through the constricting inlet, flow becomes uniform again, and Manning's equation can be used to evaluate velocity in the middle of the pipe.

The chart on the next page applies to a more or less horizontal culvert with a squared-off entrance such as a headwall. It applies to both concrete and corrugated metal (c.m.) culverts. Moderate differences in culvert slope have a negligible effect on inlet control capacity. Enter the chart from the bottom with the depth of ponded water above the culvert's invert. Move up to the necessary Q. If you fall directly on a curve for one of the pipe diameters, that is a suitable diameter for the needed capacity. If you fall between two curves, move up to the left to find the next larger size.

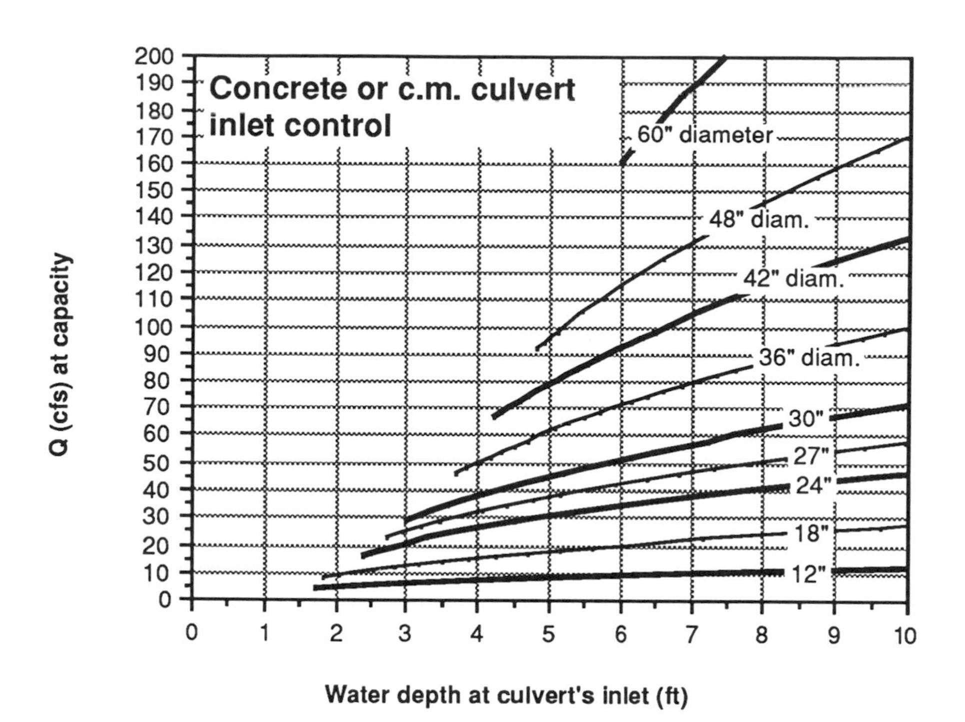

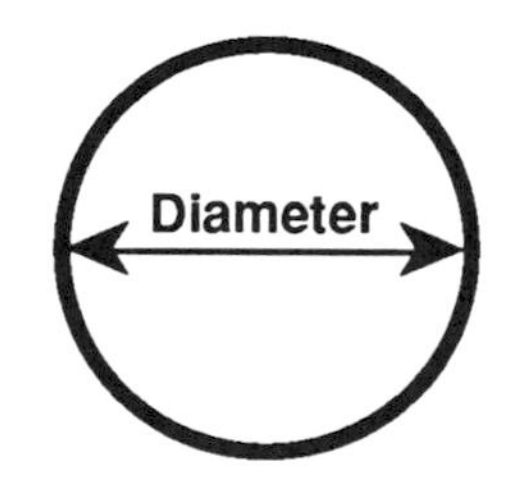

Velocity in a full-flowing circular pipe is given by,

$$V = 183\, Q \,/\, D^2$$

where,

V = velocity in feet per second (fps);
Q = rate of flow in cubic feet per second (cfs); and
D = diameter of pipe in inches.

Derived from inlet control equation for submerged circular concrete culvert at 2% slope having square edge with headwall, on pages 146 and 147 of Jerome M. Normann and others, 1985, *Hydraulic Design of Highway Culverts*, Hydraulic Design Series No. 5, Washington, D.C.: Federal Highway Administration. Velocity formula was derived from equation $Q = V A$.

Outlet Control

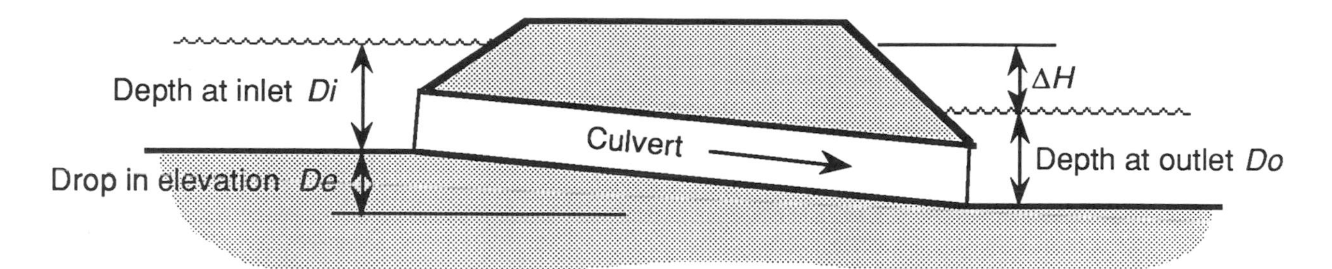

Outlet control applies where the flow of water through a culvert is controlled by depth of water at the outlet. It could occur where water is backed up at the outlet by ponding or sluggish flow. It might not occur where the water is allowed to flow freely away from the outlet. In areas of level topography such as parts of Minnesota or Florida, it is common practice to check the outlet of every culvert for its outlet-controlled capacity. Even in areas of rolling topography, outlet control can occur where pipes and swales are laid out with low gradients.

To move through the pipe water has to push out the outlet, as well as in at the inlet. The difference in elevation between headwater and tailwater establishes the energy gradient that pushes water through the pipe.

The head difference can be found from,

$$\Delta H = De + Di - Do$$

where,

ΔH = the difference in head (elevation) between the headwater and tailwater surfaces, in feet;
De = drop in elevation of culvert invert from inlet to outlet, in feet;
Di = depth of water at inlet, in feet; and
Do = depth of water at outlet, in feet.

The chart on the next page shows the ratio of rate of flow squared to head difference. For each size of pipe, the longer the pipe, the less water can be pushed through for a given head difference, since a longer pipe accumulates more friction and the gradient of the water surface going downstream declines. To use the chart to find an adequate culvert size, first derive the head difference from the above equation, and use it with your estimate of peak rate of flow to find the ratio $Q^2/\Delta H$ in cfs^2/ft. Enter the chart from the bottom with the length of the culvert, and from the side with the flow-squared-to-head ratio. If your lines intersect exactly on the curve for one of the pipe sizes, then that is your design size. If your lines intersect at a point between two of the pipe-size curves, move up and to the right to find the next larger commercially available size.

After water has passed through the constricting inlet, flow becomes uniform again, and Manning's equation can be used to evaluate velocity.

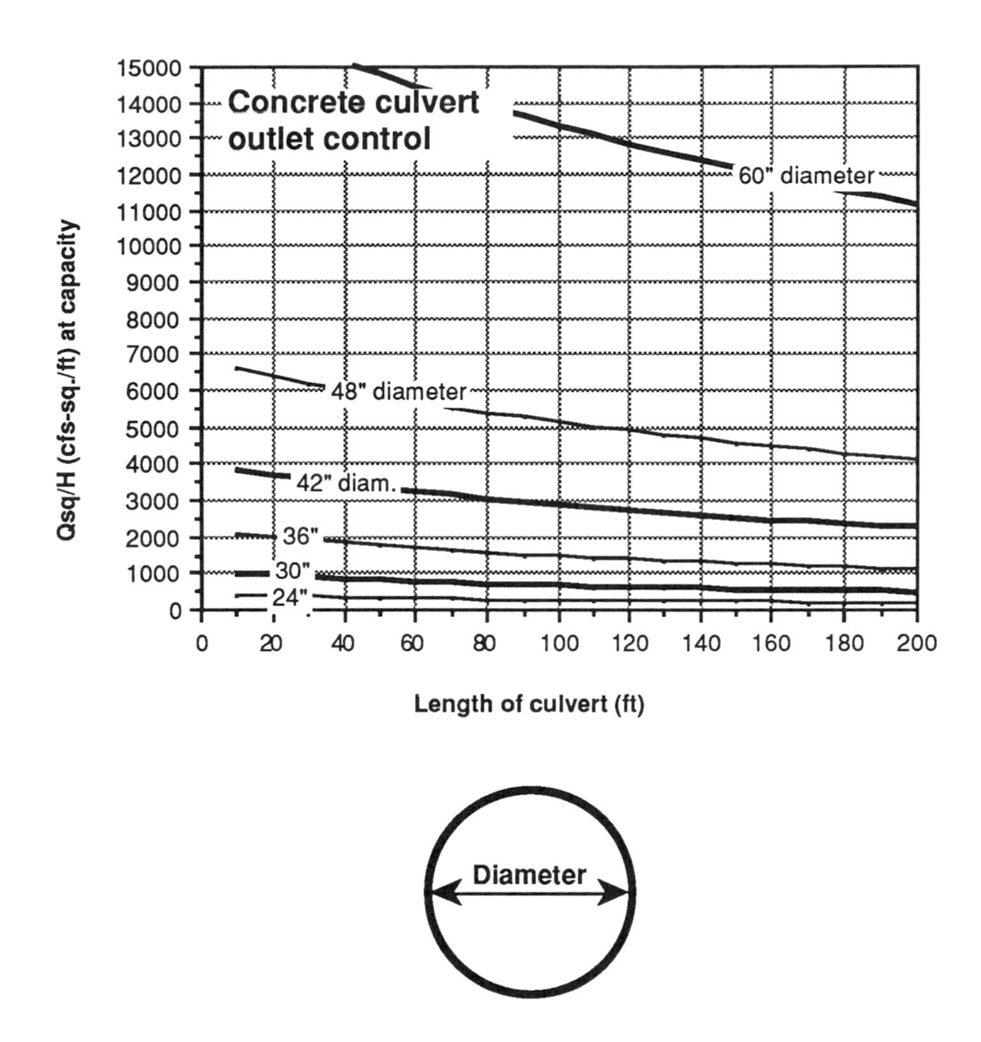

Velocity in a full-flowing circular pipe is given by,

$$V = 183\, Q / D^2$$

where,

V = velocity in feet per second (fps);
Q = rate of flow in cubic feet per second (cfs); and
D = diameter of pipe in inches.

Chart was derived from outlet control equation for culvert submerged at both inlet and outlet and flowing full with square entry (ke = 0.5 and n = 0.013), equation 5 on page 35 of Jerome M. Normann and others, 1985, *Hydraulic Design of Highway Culverts*, Hydraulic Design Series No. 5, Washington, D.C.: Federal Highway Administration. Velocity formula was derived from the equation $Q = VA$.

Weir

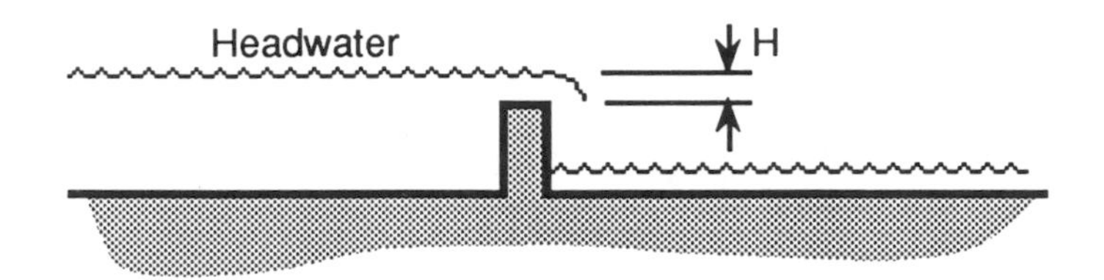

A weir is a shelf that water falls over, such as the top of a barrel pipe or the edge of a concrete wall.

Discharge is governed by the familiar principle: the higher the head upstream of the weir, the greater the flow. The type of material is irrelevant as long as the weir has only enough width for structural strength and water does not contact it long enough to be slowed down by friction.

Discharge can be predicted by the weir equation:

$$Q = C\,L\,H^{3/2}$$

where,

Q = discharge, cubic feet per second;

C = coefficient equal to 3.3 where weir has significant height on the upstream side such as at a dam or a barrel pipe, or 3.09 where the weir has no upstream height such as at a drop structure in a channel;

L = horizontal length of weir, such as the length of a straight weir in a dam or the circumference of a barrel pipe; and

H = head of water above the edge of the weir, in feet.

At a rectangular weir, the weir equation is applied directly. The chart on the next page shows discharges through various lengths of rectangular weirs using a C of 3.3. At a drop structure in a channel, the discharge would be 3.09/3.3, or 0.94, times the discharge shown on the chart.

At a circular weir, the length L is equal to the circumference of the circle, or $2\pi R$. The chart on page 98 shows discharges through commercially available diameters of pipe that might be used as barrel pipes. The chart is based on a full circle, with water free to approach from all sides of the pipe's circumference. At a half-circle weir such as a concrete weir installed as part of a dam, the discharge would equal half of the discharge shown in the chart.

A barrel pipe at a dam is often discharged under the dam through a culvert attached to the bottom of the barrel pipe. At the junction of the two pipes, an inlet or outlet control condition exists, since the water now has no forward velocity and its ability to enter the mouth of the culvert will determine how fast it can get going again. The culvert should be sized for either inlet or outlet control, whichever is more limiting.

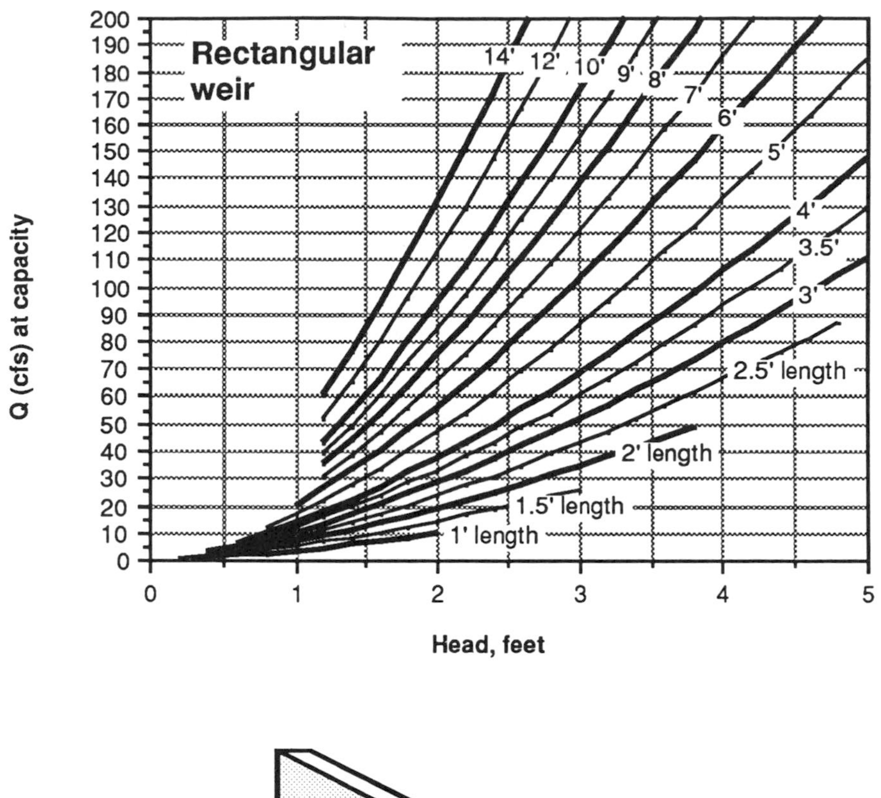

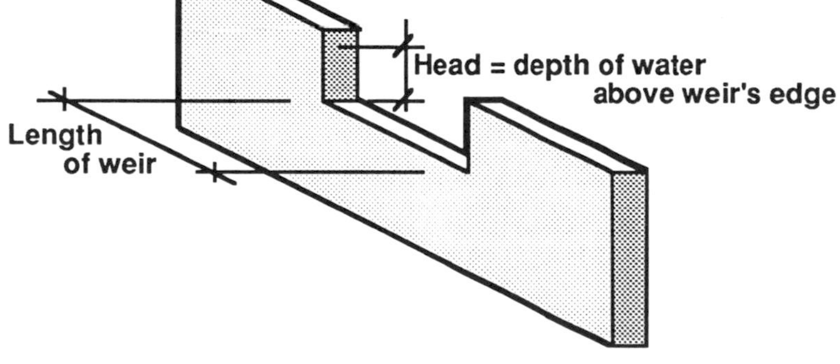

The discharge over a weir with no height above the channel, such as a drop structure, is 0.94 times the discharge given here.

Chart was derived from weir equation with $C = 3.3$. Forward velocity over the weir was assumed zero.

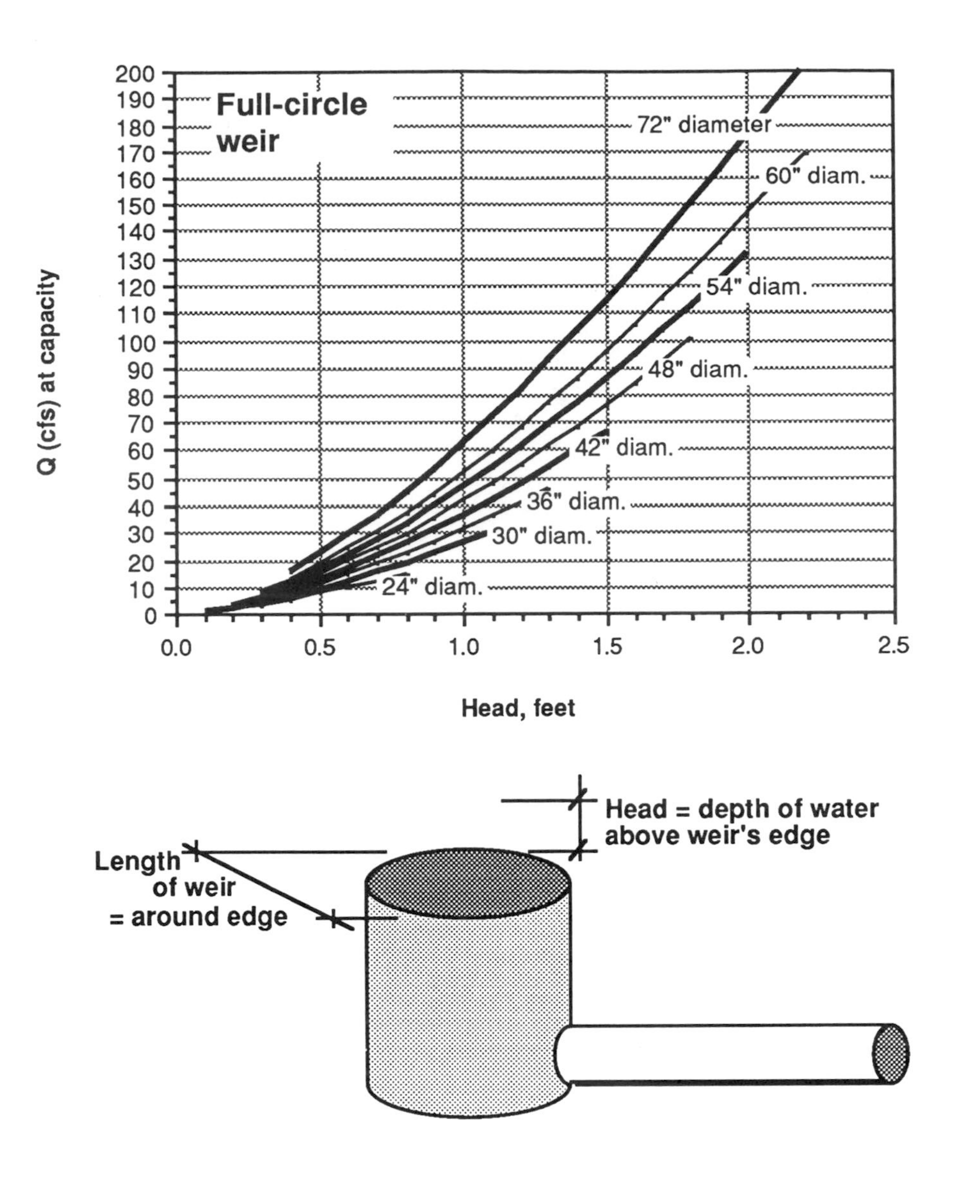

The discharge over a weir with no height above the channel, such as a drop structure, is 0.94 times the discharge given here. Discharge through a half-circle weir is half of the Q shown here.

Chart was derived from the weir equation with $C = 3.3$. Length L of a circular weir $= 2\pi R$.

Compound Conveyances

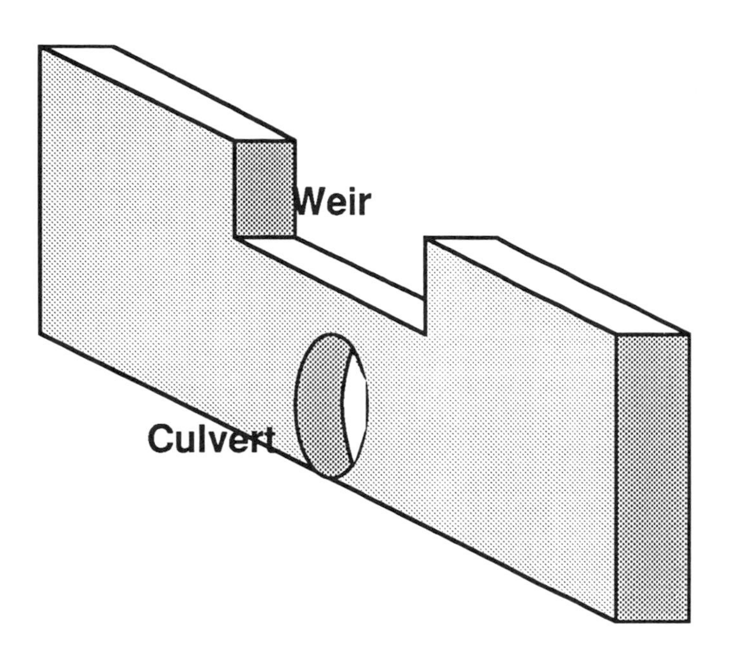

Compound conveyances can be built to discharge a combination of different design storms at controlled rates, or to direct low flows to treatment systems and high flows separately to flood detention basins.

Imagine a weir overlying a culvert. As the water level rises behind the structure, the head at the weir would be zero until the water's elevation reaches the edge of the weir. Then a positive head would begin to operate at the weir, while the total head would apply to the culvert.

Summary of Process

Define problem:

1. Identify type of conveyance from a site map or site visit. Specify whether swale, culvert or weir; cross sectional shape; type of material; slope; and any limitations on allowable headwater and tailwater depths.

2. Obtain ***Qp*** after development in cfs from previous storm runoff estimates (page 32 or 50).

Design for capacity:

3. Select a conveyance with sufficient capacity from the appropriate chart on pages 88-91, 93, 95, 97 or 98. At a culvert, if you are uncertain whether uniform flow, inlet control or outlet control applies, check all three charts to find which is the most limiting.

Check for velocity:

4. Obtain maximum permissible velocity where applicable from the table on page 82 or local regulations and practices.

5. Compute velocity by applying the equation shown at the chart for each kind of conveyance, or by the general equation,

Velocity in fps = *Qp* in cfs / Cross-sectional area in sq.ft.

6. If velocity exceeds the maximum permissible velocity select a conveyance with larger capacity, armor the erodible soil with a stable material, or revise the site plan to reduce conveyance gradient or rate of runoff reaching the conveyance, and begin again with step number 1.

Conveyance Exercise
1. Triangular swale, after development

		Site 1	*Site 2*
Define problem:			
Swale material (turf, concr. or riprap)	=	____	____
Swale gradient (from site plan)	=	____ %	____ %
Max. permitted flow depth (from site plan)	=	____ ft	____ ft
Max. noneroding velocity (from page 82)	=	____ fps	____ fps
Design based on rational method:			
Qp by rational method (from page 32)	=	____ cfs	____ cfs
Flow depth *D* (from page 90)	=	____ ft	____ ft
Velocity $= Qp/(10D^2)$	=	____ fps	____ fps
Design based on SCS method:			
Qp by SCS method (from page 50)	=	____ cfs	____ cfs
Flow depth *D* (from page 90)	=	____ ft	____ ft
Velocity $= Qp/(10D^2)$	=	____ fps	____ fps

Conveyance Exercise

2. Circular concrete pipe, after development

		Site 1	*Site 2*
Define problem:			
Qp by SCS method (from page 50)	=	cfs	cfs
Max. noneroding velocity at outlet (from page 82)	=	fps	fps
Design based on Manning's equation:			
Culvert diameter D (from page 88)	=	in	in
Velocity through culvert $= 183Qp/D^2$	=	fps	fps
Design based on outlet control:			
Slope of culvert (from site plan)	=	ft/ft	ft/ft
Length of culvert (from site map)	=	ft	ft
Loss of invert elev. in pipe De = (culvert length)(slope)	=	ft	ft
Allowable depth at mouth Di (from site map)	=	ft	ft
Depth of flow at outlet Do (from page 101)	=	ft	ft
Allowable head loss ΔH $= De + Di - Do$	=	ft	ft
Ratio Q^2/H $= Qp^2/\Delta H$	=		
Culvert diameter D (from page 88)	=	in.	in.
Velocity through culvert $= 183Qp/D^2$	=	fps	fps

Conveyance Exercise
Discussion of Results

1. Which site requires the larger pipe or deeper swale? Why?

2. Is peak rate of flow in cfs the only consideration that may limit the size of a conveyance? Why? In what specific types of site conditions could other considerations limit conveyance size?

3. Which runoff estimation method requires the larger swale? Why? Deeper, broader swales preempt more valuable urban land. What conveyance sizing procedure would you use in practice? Why?

4. Which culvert sizing method requires the larger culvert? Why? Larger pipes cost more money to build. Which size is the *correct* size? Why? Why did the other method *fail* to yield the correct pipe size? Which sizing method would you use in practice? Why?

5. What are some approaches you might take to reducing velocity in a conveyance? If in practice reducing the velocity in a conveyance seems impractical, what are some other approaches you could take to reducing channel erosion?

6. Some textbooks present charts based on Manning's equation as ways to determine a pipe's capacity. Under what conditions can such charts be used that way? What circumstances might make the exclusive use of such charts invalid? Where their use is invalid, what alternative approach should you take?

Summary and Commentary

Conveyance is a *design objective*. It is a hydraulic function for a site. It is a scenario for a *kind* of way water moves through a site.

Conveyance addresses on-site nuisance and safety. It does *not* address other water resource or environmental concerns such as water quality, flooding, stream base flows, downstream channel erosion, or ground water.

A variety of conditions might limit the rate at which water flows through a conveyance. Each conveyance should be checked for all the possible conditions that may limit rate of flow through it.

Existing swales and culverts are sometimes included as parts of the conveyance systems of new developments. When this occurs, you can evaluate them by using a capacity chart in reverse. On an appropriate chart enter the known characteristics of such a conveyance and find the flow and velocity that it can carry. Compare your results with the estimated peak flow of runoff that will enter it to see whether any reconstruction is necessary.

Hydraulics need not dictate urban design. The results of conveyance calculations are a pipe's or swale's size, material, gradient and velocity. These data are *parameters for site design.* As long as you meet these few criteria, the hydraulic function will work, and you are free to *design the site* any way that is needed in terms of planting, contouring, additional materials etc.

Chapter 5

Storm Detention Design

Detention is design for slowing down the rate of flow as stormwater runs over the ground.

Storm detention is delaying of flows during storm events in order to reduce flood peaks.

Urban storm detention started in the 1960s when it was discovered that development tends to be followed by an increase in *Qp* for a given storm. Impervious roofs and pavements tend to increase the volume of rainfall that becomes runoff, and shorten runoff's travel time. The new combination of volume and velocity can cause great increases in peak storm flow downstream.

Aggravated flood damage is the result. The homes ruined, the industries closed, the mud saturating farmers' fields, the roads closed off and the bridges collapsed, have motivated many local and state governments to require on-site detention. Although it has too often been poorly regulated, and the facilities for it poorly implemented and poorly maintained, storm detention is motivated by a real concern for people, property and the environment.

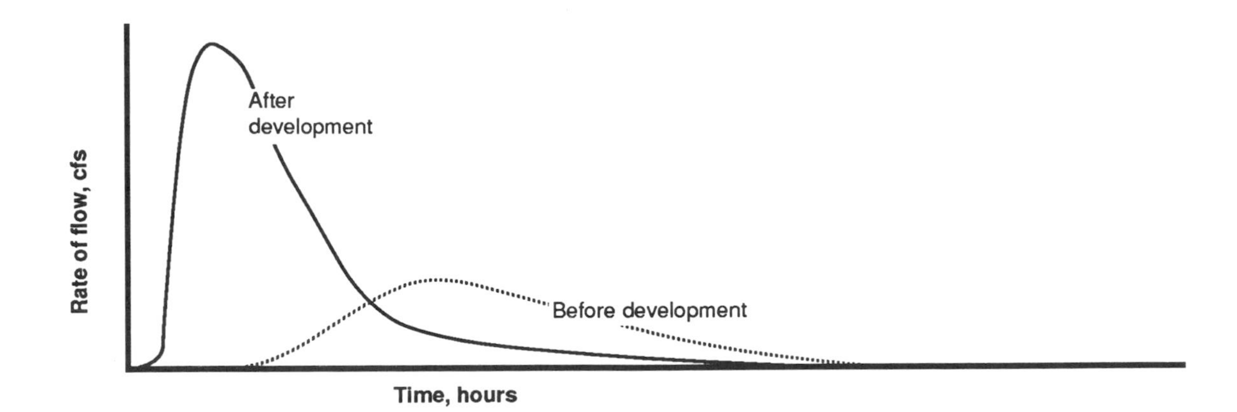

Peak rate of runoff is the target of storm detention.

One or more points on the stream system need to be defined where the benefit of detention is to be effected. This is often a point where a major swale or stream leaves a development site, or a point farther downstream where a combination of on-site and off-site flows can be evaluated.

A common goal for detention is that the *Qp after* development should not exceed that for the same drainage area *before* development. The *Qp* before development is estimated using conventional runoff models. In some regions the goal is that *Qp* not exceed that coming from the same drainage area under a theoretical standard land use such as a meadow, which can also be estimated using the same runoff models.

Adding a detention facility to a development does not change the fact that an impact is created. It *insulates* outside areas from the impact that is created on site. By controlling peak outflow from a developed area it makes rivers downstream "think" that the development never really happened (at least as far as the single variable *Qp* is concerned).

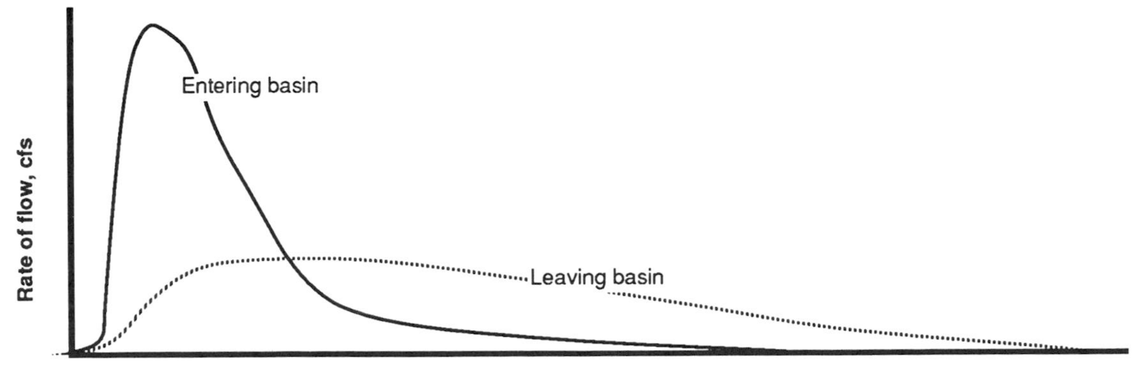

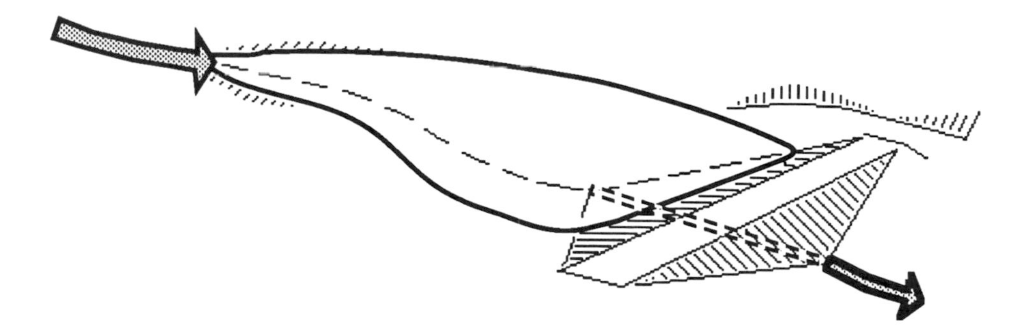

A basin is the basic facility to implement detention by temporarily storing a volume of water during a storm event. A constricted outlet retards the basin's rate of outflow. Stormwater entering with a high *Qp* backs up behind the outlet. The idea is to make the outlet small enough to reduce the flow to the required rate, and the basin big enough to store the difference.

The flood storage volume is *above* the invert elevation of the basin's outlet. This is where runoff will accumulate during the storm event, to be let out relatively gradually through the deliberately restricted outlet. This flood-control volume sits in reserve, full only of air, almost all the time. It is used for holding water only during flood events.

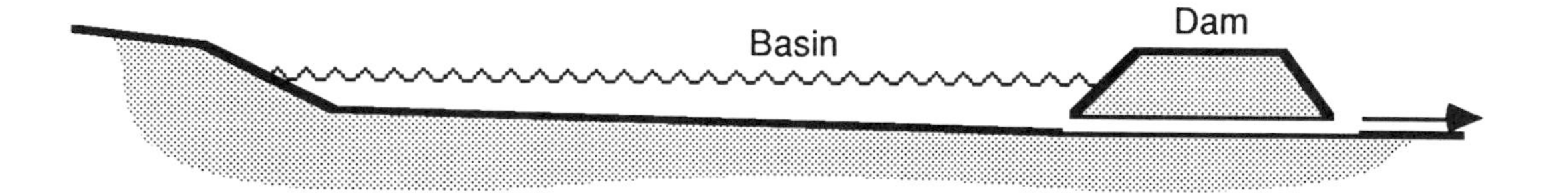

In a dry basin the outlet's invert is flush with the basin's floor. When water backs up in the basin the floor forms the bottom of the storage volume.

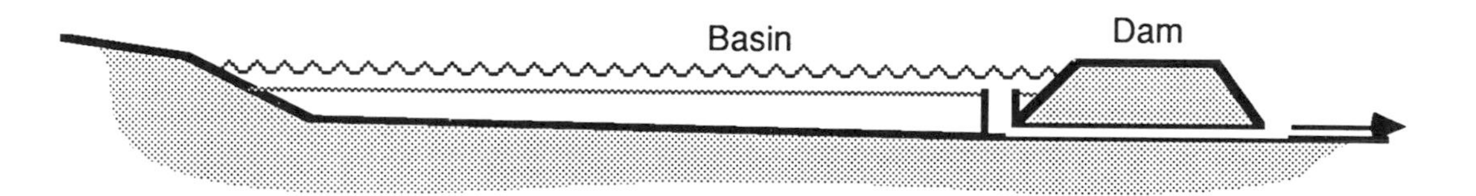

In a wet basin a raised outlet establishes the water level of a permanent pool. When storm runoff enters the basin the additional volume rests temporarily on top of the permanent water surface. The volume of water permanently in the pool is not counted in storm detention storage, since it is already preempted.

There are many ways to design storm detention basins into the landscapes of site developments. Choices of location, form and materials are a matter of site design for aesthetics, multiple use, wildlife, safety, maintenance and cost, as well as hydrologic objectives. A wet pond can be shaped as an amenity. Attempts have been made to get multiple use out of the floors of dry basins, but wet-soil vegetation and occasionally ponded water are constraints that must be overcome.

In Canton, Massachusetts a dry basin has been squeezed into a few hundred square feet in a townhouse-apartment development. The basin is bounded on one side by a stone-veneered retaining wall and on the other sides by steep slopes covered with cobbles. The basin is surrounded by grass and a canopy of native maple trees, and is overlooked by several residential buildings. Two culverts drain into the basin; the outlet is a third culvert that passes under the road that bounds the property. There is often a small throughflow from the surrounding soil. Within a couple of years after construction was completed sediment carried in by stormwater began to fill in low points among the cobbles. Grasses and lowland plants are gradually growing up along the basin's narrow floor.

Near Columbus, Ohio two wet basins at a suburban industrial headquarters were designed by James H. Bassett as amenities. One pond has very little detention storage capacity. It is located directly outside the lobby, cafeteria and other public rooms of the building and is ornamented with patios, paths, plantings and lighting. The second pond's major function is stormwater storage; it has gravel edges to accommodate water-level fluctuations. It is located farther from the building and lower in elevation, and is surrounded by large grass areas and groves of trees compatible with the remainder of the grounds and with long vistas of the headquarters building across the large expanse of water. The ponds are connected by a rocky, carefully planted artificial waterfall supplied by a pump that recirculates water from the lower pond to the upper. If the water in the upper pond ever falls below the level needed for amenity it can be replenished with city water.

In Rio Rancho, New Mexico a dry basin in a residential development is turfed, irrigated and ornamented at the upper elevations with trees, lighting and play equipment. A concrete low-flow channel through the basin has gentle asymmetrical slopes so that it doubles as an actively used walkway. The dam's concrete overflow channel is equipped and marked as a basketball play area.

A secondary overflow is as necessary for a basin's outlet as for any other conveyance. A secondary overflow, or *emergency spillway*, must pass flows larger than the design storm without eroding the dam or other critical parts of a basin. Some common forms of secondary outlets are earthen channels excavated at the side of the dam, and large weirs or pipes at elevations higher than the principal outlet. Where the principal outlet is designed both to control the design storm and to pass larger flows, a separate emergency outlet of limited capacity is not necesary.

Protection from erosion is as necessary for a basins's outlet as for any other conveyance. Velocity at the outlet should be evaluated and compared to the non-eroding velocities listed on page 82. If necessary, it shoud be reduced or compensated for as discussed in the chapter on *Conveyance Design.*

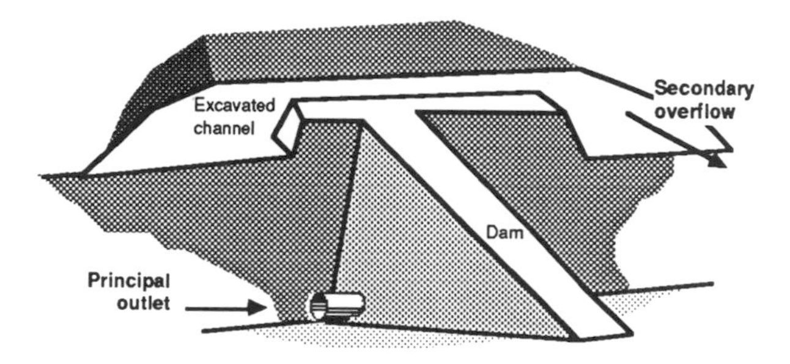

Water quality is improved by storm dctention only to a small degree. Near Washington, D.C., trace metals such as lead and zinc were found to have accumulated in the surface soils of dry basins. The metals were mostly in a non-leachable form, and almost no downward movement of the metals through the soils had occurred. Accumulation seemed to be a function of micro-topography in the basin and the resultant residence time of standing water. The concentrations of metals were below any level that might be considered harmful to the soils or to the surrounding environment. Downstream channel erosion and resulting turbidity may be *increased* by storm detention, when detention storage prolongs the time during which storm flow is above a threshold velocity for channel erosion. If you want significant surface water quality improvement, you have to design explicitly for it, using the procedures described in the chapter on *Extended Detention Design.*

Ground water and stream base flows are similarly affected by storm detention only to a small degree. Although some water infiltrates during detention, infiltration is ordinarily very slow compared with surface outflow. It is safest to design detention with the assumption that infiltration will be zero during the detention period. If you want a significant proportion of flows occurring over a long period of time to go into the soil, you have to design explicitly for it using the procedures described in the chapter on *Infiltration Design.*

Data on Washington-area basins are from Parker J. Wigginton and others, 1983, Accumulation of Selected Trace Metals in Soils of Urban Runoff Detention Basins, *Water Resources Bulletin* vol. 19, no. 5, pages 709-718. Data on stream-channel erosion following detention are from Richard H. McCuen, 1987, Multicriterion Stormwater Management Methods, *Journal of Water Resources Planning and Management* vol. 114, no. 4, pages 414-431.

Stage-storage-discharge is the basic relationship governing both storage volume and outflow rate. This is a site-specific graph or table showing how a basin's storage volume and outflow rate vary with changes in water elevation.

Stage is the height of the water surface above the outlet's invert elevation or another base datum. Stage increases during a storm, and subsides to its original level afterwards.

Storage is the volume of flood water held in a basin. As stage increases, storage also increases. A stage-storage curve can be drawn by estimating, from a topographic plan, the volume that a basin can hold at a number of given stages. The maximum storage during the design storm must be adequately provided for in the basin's design.

Discharge is the rate of outflow through the outlet. As shown in Chapter 4, the rate of flow through a conveyance depends on the height to which water rises. As stage increases, discharge also increases. A stage-discharge curve can be drawn by reading the discharge rates at a number of given water depths from an appropriate chart in Chapter 4. A basin's design must limit the maximum discharge during the design storm to an allowable rate.

Since storage and discharge are each related to stage, they are related to each other. As storage increases, discharge increases. The maximum discharge occurs simultaneously with the maximum storage. That maximum moment is the subject of detention design.

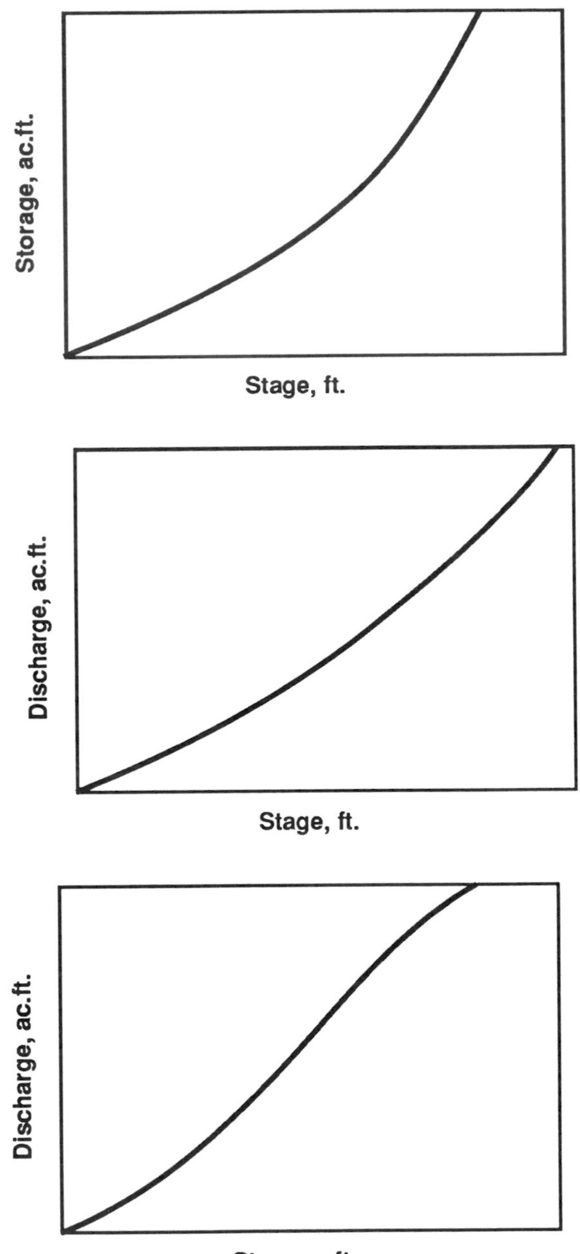

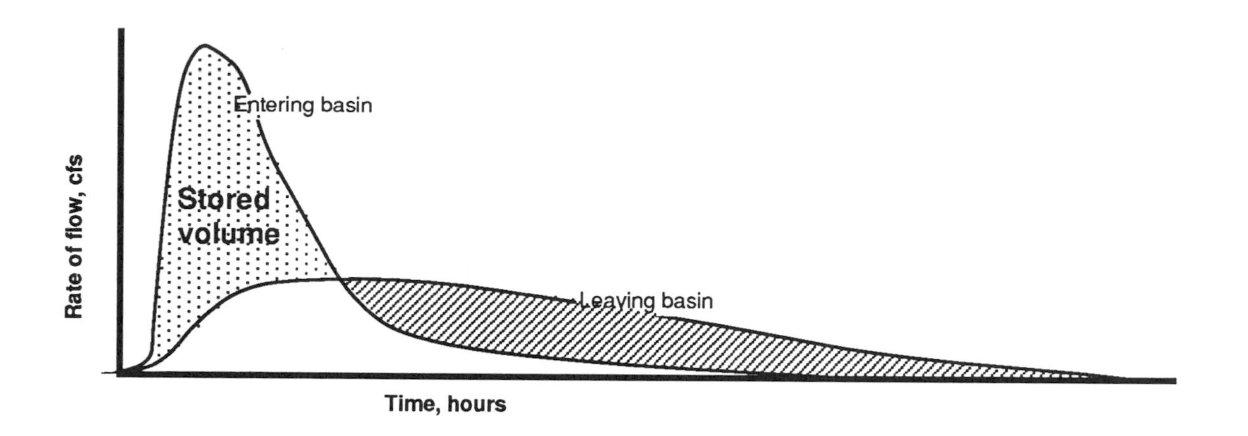

Derivation of required storage volume in a basin is based on an application of the basic reservoir equation. For any increment of time,

$$\Delta \text{Storage} = \text{Inflow} - \text{Outflow}$$

In the illustration one curve shows the flow *entering* a basin from a developed site. The second curve shows the flow of the *same* water *leaving* the basin. In the early (left) part of the graph, outflow starts almost as soon as inflow, but the outflow rate is lower because of the deliberately constricted outlet. The difference between inflow and outflow is the water that has entered but not yet left. It is accumulating in the basin. The total volume that accumulates is equal to the area in the hydrograph between the inflow and outflow curves. That is the required capacity of the basin.

Reservoir routing is a thorough, precise way to estimate required storage volume, in which inflow, outflow and storage are computed for short increments of time and accumulated over the duration of the storm event. When done by hand, it can be a complicated and time-consuming procedure.

Computer programs that include reservoir-routing routines can do site-specific routing calculations with ease, requiring only a stage-storage-discharge table from the user and a runoff hydrograph to route through it. The disk which accompanies this book is one of the programs that combines these capabilities. Its operation is described in Appendix H. Such programs can be used in a trial-and-error design process, in which simulated flows are routed through proposed basins to see whether a basin can hold a required volume.

Without a computer to do reservoir routing, adequate estimates of required storage volume in small basins can be made based on previous experience with calculations for other basins.

The chart on the next page is based on two ratios which are convenient ways of generalizing about the hydrologic performance of detention basins. The chart can be easy to use as long as you spend a minute getting a feel for the intuitively reasonable principles on which it is based.

The ratio of peak flow leaving to peak flow entering a detention basin indicates how much "work" the basin is doing. *Qp* into a basin is the flow from a proposed development; *Qp* out is the maximum rate at which water leaves the basin and continues downstream. If the ratio is low (approaching 0) *Qp* is reduced greatly; it is intuitively reasonable that a large basin would be required to take up the large difference in flow. If the ratio is high (approaching 1) *Qp* is reduced little; it is intuitively reasonable that only a small basin would be required.

The ratio of storage volume to runoff volume (*Vs/Qvol*) indicates how big the volume of a basin is in relation to the total volume of runoff that passes through it during a design storm. A basin capable of doing lots of detention work holds a large portion of the runoff volume (*Vs/Qvol* approaching 1). A basin without so much work to do could be smaller in relation to the design storm (*Vs/Qvol* approaching 0).

Thus if we know how much we want to suppress the peak flow (*Qp* out/*Qp* in), then we ought to be able to derive how much of the design storm volume needs to be stored (*Vs/Qvol*).

The curves on the chart show relationships between the two ratios found by SCS in numerous computer reservoir routings. There are separate curves for different storm types, since timing of flow during a storm results in different storage regimes. At the upper left corner of the chart a basin has a lot of work to do (peak outflow/peak inflow approaching 0) so the basin needs to be large (high storage volume/runoff volume). At the lower right corner runoff is discharged quickly relative to inflow, so the basin need not store a large part of the storm runoff volume.

To use the chart for design, first find the desired ratio of *Qp* out to *Qp* in from storm runoff estimates and local detention standards. Then enter the chart from the bottom at the desired ratio. Move up the chart to the appropriate curve for your rainfall region (see the map on page 44), thence horizontally to the scale on the left, where you can read the required ratio of storage volume to runoff volume (*Vs/Qvol*). The volume of runoff *Qvol* entering the basin was found in your runoff estimate on page 50. The required storage capacity *Vs* of the basin is then:

$$Vs = (Vs / Qvol) Qvol$$

To check the basin size indicated by the chart you can use a reservoir routing procedure on a computer program.

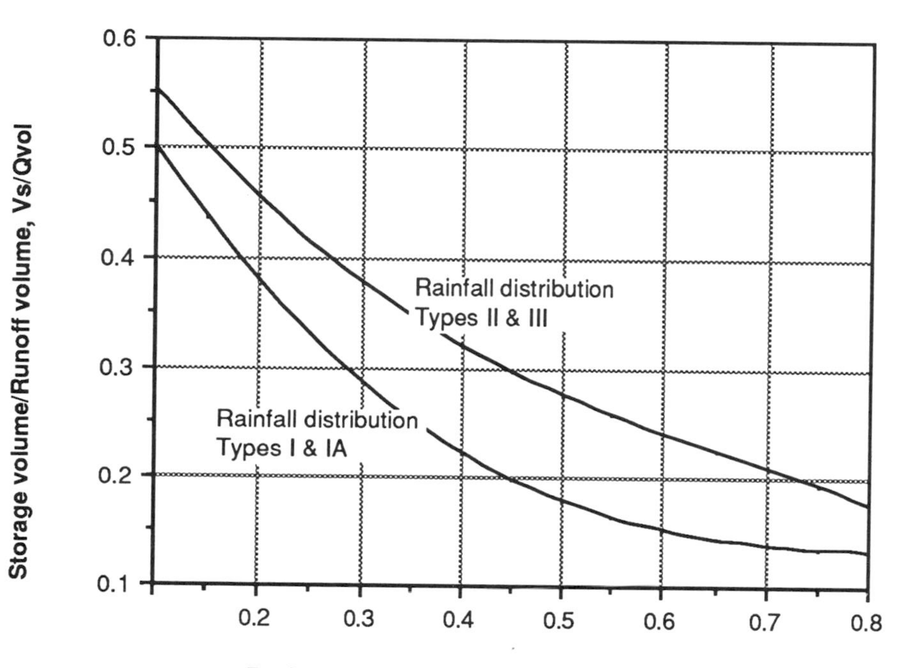

Chart was derived from equation for Figure 6-1, given on page F-1 of U.S. Soil Conservation Service, 1986, *Urban Hydrology for Small Watersheds*, Technical Release 55, second edition.

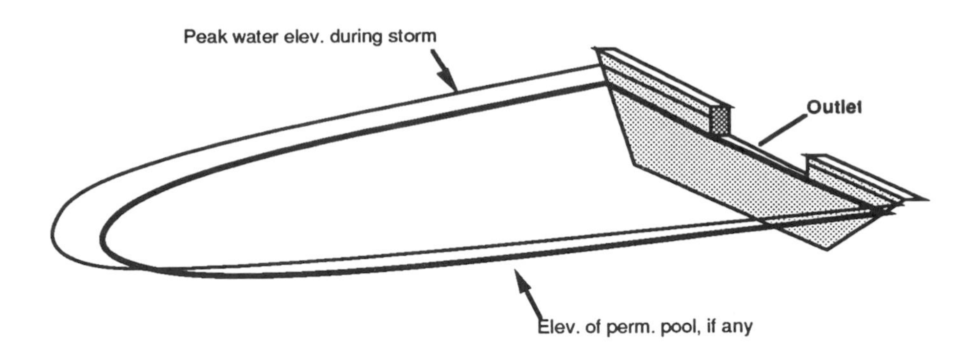

The exact layout in plan of a basin to meet a given storage requirement depends on site-specific constraints and creative landscape design. On a cramped site you might be forced to make a basin deep and narrow to fit the available area. On a more generous site you might make it broad and shallow to blend with surrounding landforms and allow multiple use.

The volume must fit between the elevations of the outlet's invert (at the bottom) and the maximum water level during the design storm (at the top). Complete all contouring and spot elevations necessary to show that your basin satisfies the volume requirement. You can estimate the volume of a basin from the plan using the same volumetric methods that are used for earthwork, such as contour-planes. In this case you are finding the volume of "air" in the basin, not the volume of earth around it. For further information about basin grading see Appendix C.

Note the depth (stage) of water that you end up with at the basin's outlet. The depth is the difference in elevation between the outlet's invert, and the maximum level of stored water. Depth will push water into the outlet, so for appropriately sizing the outlet we need to know depth at the moment of maximum storage.

Select an outlet from the charts in Chapter 4 that will provide the intended outflow rate at the given stage. A detention outlet deliberately slows down runoff. Manning's assumption of free and continuous flow does not fit. The outlet should be considered a weir, or a culvert with either inlet or outlet control.

If your outlet is a rectangular weir, you can interpolate between the curves on the weir chart to derive a precise required length. Rectangular weirs tend to be custom-built of concrete or masonry, and can fit any specified size.

If your outlet is a culvert or a circular weir made from a culvert, your choices are restricted to commercially available sizes. If you fall between two of the curves for culvert sizes, then:

a) Choose a smaller culvert size, and read its outflow rate at the given depth. Recompute the basin's volume requirement using a new ratio of *Qp* out to *Qp* in and revise the basin's grading plan as needed. Alternatively,

b) Choose a smaller size, with the given outflow rate *Q* corresponding to a greater depth than originally planned. The required depth, if not excessively greater than originally planned, can be produced in the basin by dropping the invert elevation or raising the water level, and recontouring the rest of the basin to fit.

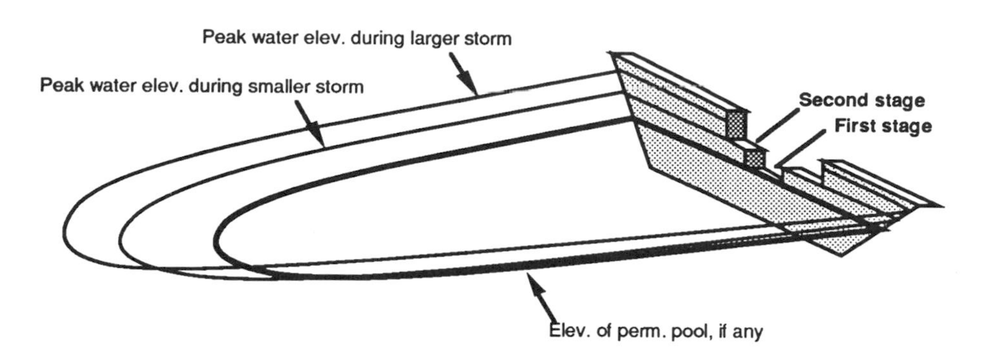

Two or more design storms are required for flood control in some areas. Here are two approaches to meeting such criteria:

1. Design the volume and outlet first for the largest storm. Develop a stage-storage-discharge curve from the appropriate conveyance chart and the basin's grading plan. Then use the volume chart on page 111 or a reservoir-routing computer program to determine stage, volume and outlet rate that will occur when a *smaller* storm comes through. Since the depth at the outlet will be lower during smaller storms, the outflow rate will be lower. In some cases it happens that an outlet designed for the largest storm also controls smaller storms adequately. If not, possibly reducing the size of the outlet for the largest storm, thereby overcontrolling the largest storm, could create a basin that works well for smaller storms as well. The storm that requires the smallest outlet is the one that sets the design size. Any reduction in outlet size is accomapanied by greater stage and storage in the basin, since outflow rate is reduced.

2. Design a multiple-stage outlet. Select a basin location with a large volume capacity, and develop a stage-storage curve for it. Select an outlet for the smallest storm using the appropriate conveyance chart. At the stage to which water will rise during the smallest storm, set the invert elevation of the outlet's second stage. Use the volume chart on page 111 to find the storage, and thus elevation, to which water will rise during the larger storm. The head *during the larger storm* over the first (lower) stage of the outlet is given by,

$$\text{Head} = \text{Water elevation} - \text{Outlet invert elevation}$$

The flow through the first stage during the larger storm can then be found from the appropriate conveyance chart. The remaining flow to be released by the upper stage of the outlet is,

$$\text{Remaining } Q = \text{Total } Q - Q \text{ through first stage.}$$

You can then select a second stage of the outlet to pass this rate of flow.

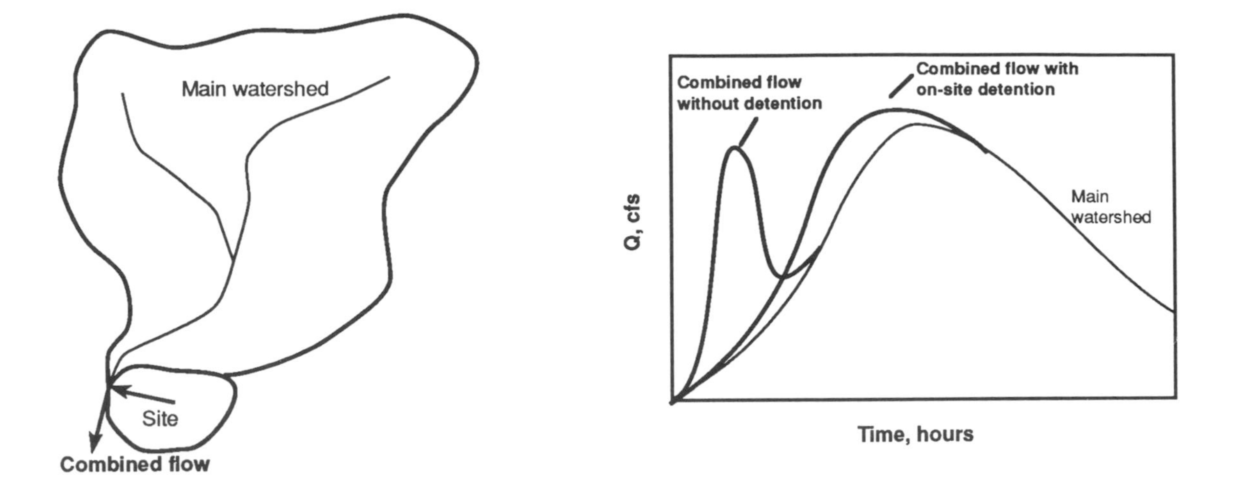

Timing can make a difference.

Detention can sometimes aggravate floods downstream, because temporary storage *delays* a site's peak flow, in addition to *reducing* it.

Imagine a small site that drains into a stream with a watershed much larger than the site. The storm hydrograph from the main watershed is long and slow, and the peak flow late, compared to the flow from the site. Since the site's flow drains out before the main watershed's peak arrives, it does not contribute to the magnitude of a flood downstream. But if detention is added to the site, outflow from the site will be delayed, so that it may combine with the peak flow on the main watershed and contribute to the peak flow downstream.

Also imagine two detention basins on different sites in the same watershed, constructed by different developers at about the same time. When the hydrographs from the two basins combine downstream, their delayed flows combine in a way that had never existed before development, and a larger flood may be created.

The only practical way to check for downstream effects of timing is to use a computer program that routes flood hydrographs through the drainage system over time. Take your analysis as far downstream as the effects of your development might significantly extend. The disk that accompanies this book can do such an analysis. After you input all the watershed, channel, and reservoir conditions and specify the rainfall, it routes the hydrographs from separate sub-watersheds through the drainage network, so you can see the results of your proposed plans.

The idea for on-site detention basins came from regional flood control reservoirs, which were successfully implemented by public agencies on major rivers in the early part of this century. The concept of on-site basins transfers responsibility for flood control to developers whose projects aggravate flood peaks, and disperses its implementation to numerous small facilities throughout river basins. A flood-control detention requirement can do more harm than good if it is applied uniformly, without regard to site-specific downstream impact.

Summary of Process
Storm detention

Volume of flood storage:

1. Specify detention goal: peak rate of outflow in cfs desired after development during a selected design storm. If this is the rate before development use previous storm runoff estimate on page 49.

2. Obtain the peak rate of flow that enters the basin after development (Qp), in cfs, from storm runoff estimate on page 50.

3. Compute the peak flow ratio (no units) from the results of steps 1 and 2:

$$Qp \text{ out} / Qp \text{ in} = \text{Peak flow out of basin / Peak flow into basin}$$

4. Read the volume ratio (no units) from the chart on page 113:

$$Vs / Qvol = \text{Volume of storage/Volume of runoff}$$

5. Obtain the volume of runoff ($Qvol$) in ac.ft. entering the basin after development, from your storm runoff estimate on page 50.

6. Compute required storage volume (Vs) in acre feet from the results of steps 4 and 5:

$$Vs = (Vs/Qvol)\,(Qvol)$$

Outlet:

7. Lay out the basin physically on the site development plan in such a way as to satisfy the storage volume requirement.

8. Read water depth above outlet's invert in the proposed basin, in feet.

9. Select outlet that will limit Qp to the desired rate at the given water depth, from an appropriate conveyance chart in Chapter 4.

10. If a precise outlet size cannot be selected because of restrictions to commercially available sizes, adjust the basin depth or volume as necessary to make the combination of depth, volume and discharge all match at the selected outlet size.

11. Check effectiveness of design by routing the design storm through your basin and any significant downstream areas using a computer program, if available.

Storm Detention Exercise

Qp after development ≤ Qp before development

		Site 1	Site 2
Volume of flood storage:			
Qp before development (from page 49)	=	cfs	cfs
Qp after development (from page 50)	=	cfs	cfs
Ratio *Qp* out/*Qp* into basin = *Qp* before/*Qp* after	=		
Required ratio *Vs*/*Qvol* (from page 113)	=		
Qvol after development (from page 50)	=	ac.ft.	ac.ft.
Required storage volume *Vs* = (*Vs*/*Qvol*)(*Qvol*)	=	ac.ft.	ac.ft.
Concrete culvert outlet (initial design):			
Max. *Q* through culvert = *Qp* before development	=	cfs	cfs
Water depth at outlet (from site map)	=	ft.	ft.
Culvert diameter with inlet control (from page 93)	=	in.	in.
Approximate area of basin = *Vs* / depth	=	ac.	ac.
Alternative 1: increase depth to match specified Qp:			
Actual depth, at *Qp* through selected culvert, if greater than initial depth (page 93)	=	ft.	ft.
Approximate area of basin = *Vs* / depth	=	ac.	ac.
Alternative 2: increase volume to match smaller culvert:			
Qp through selected culvert at initial depth, if lower than initial *Qp* out (page 93)	=	cfs	cfs
Ratio *Qp* out/*Qp* in (calculated)	=		
Required ratio *Vs*/*Qvol* (from page 113)	=		
Required storage volume *Vs* = (*Vs*/*Qvol*)(*Qvol*)	=	ac.ft.	ac.ft.
Approximate area of basin = *Vs* / depth	=	ac.	ac.

Detention Exercise

Discussion of results

1. Which site requires the larger detention basin? What factors contributed to the requirement of such a large basin? Is a great amount of *natural* (before-development) runoff sufficient cause for a relatively large basin? Why?

2. What is the approximate proportion of each site that must be occupied by a detention basin (basin area ÷ total site area)? In your judgment does either basin occupy *too much land* to be reasonable in terms purely of land use?

3. If your response to question number 2 was yes, some approaches you might take to solving the problem include changing the development's type of land use, choosing another site for this type of development, using underground or roof-top storage basins, reducing the area of impervious surfaces, modifying the quantitative standard for detention control, and going to some other type of hydrologic function. Which approach would you most likely choose? What specific site features would you have to build in order to implement your approach?

4. If your basin's outlet includes a culvert discharging into an earthen swale or stream channel, estimate the velocity there when *Qp* out of your basin passes through, using the formula by the conveyance chart that you used for your culvert. Compare your estimated velocity with the maximum noneroding velocities listed on page 82. If velocity is excessive, what would you do about it? Does taking care of excessive velocity at this location necessarily eliminate stream erosion farther downstream?

5. On-site stormwater basins are often located on rather small drainage areas. A detention basin reduces *Qp* on the small stream or swale on which it is located. But it has also *delayed* the time to peak, as shown in the hydrograph on page 104. Let us say that, a short distance downstream, your small stream drains into a much larger stream, having a longer natural time to peak. What effect if any would your basin tend to have on the total *Qp* in the major stream?

6. Taking into account the potential effect of combining flows over time mentioned in question number 5, what constraints do you think should be placed on implementing detention on a regional or river-basin scale? If you were an administrator of such a regional program, how would you make sure that the right thing is being done on each site as it is developed?

7. The peak outflow from a detention basin always coincides with a point on the receding limb of the inflow curve. Derive this principle algebraically from the basic reservoir equation,

$$\Delta\text{Storage} = \text{Inflow} - \text{Outflow}.$$

Summary and Commentary

Storm detention is another scenario for how water could move through a development site, an alternative to conveyance. It is aimed at the objective of flood control. To choose it is to choose a type of hydrologic function, a qualitative way for water to flow through a site.

Detention controls *Qp* from a development. Detention is a slowing down of the *rate* of flow in cfs. It does not affect the total flow volume (*Qvol*). The entire volume eventually outlets from a detention basin and continues to be conveyed downstream. Detention affects only individual stormwater flow events; few on-site basins are large enough to affect stream "base" flows by storing and releasing water from month to month. Storm detention does not significantly address water quality or ground water.

Downstream from a detention basin the hydrologic function returns to conveyance. Most detention projects are in fact combinations of detention and conveyance: conveyance from upstream draining into a basin, detention at the basin, and again conveyance downstream.

A detention basin is a new facility in a site development program. It must be given adequate space in a site plan, and a positive landscape design.

The results of hydrologic calculations for storm detention give the required size of basin, and associated type and size of outlet. These few specifications are parameters for site design. As long as you meet these few criteria you are free to *design the site* any way that is needed in terms of materials, planting, contouring, etc.

Although basins are frequently located downstream from site construction near a property boundary, they could with a little extra computational effort be moved upstream and integrated into the site's land uses, giving them amenity, accessibility and maintainability, and avoiding further disturbance of lowland environments. An oversized basin can compensate for the uncontrolled runoff from construction downslope. The effect of detention is evaluated downstream, where controlled and uncontrolled drainages combine.

Alternatively the detention function could be removed from individual development sites, since numerous dispersed basins can create haphazard effects on flow timing in main streams and become public maintenance nuisances. In some areas developers pay an "impact fee" rather than build an on-site basin. The fee goes to the county or other agency, which then builds regional flood control reservoirs or other stormwater management projects at selected points in river systems. Regional reservoirs are sometimes designed as public parks and can be maintained by the park agency. However, such a regionalized program sacrifices the unprotected headwaters of streams to increased urban runoff, and the total volume of runoff is still unreduced.

Chapter 6

Extended Detention Design

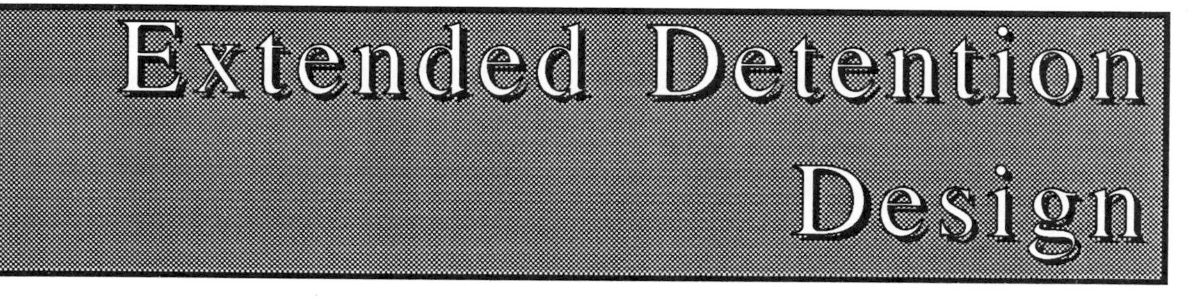

Improving water quality is the aim of extended detention.

In the still water of a permanent pool, suspended particles and the pollutants attached to them can settle out. This requires a longer storage time than flood-control detention, so detention for water quality has become known as *extended detention.* Since it requires a permanent pool, it is also known in some areas as *wet detention.*

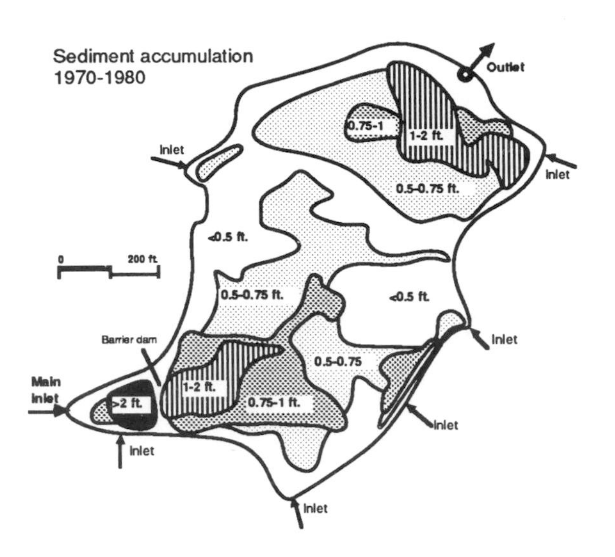

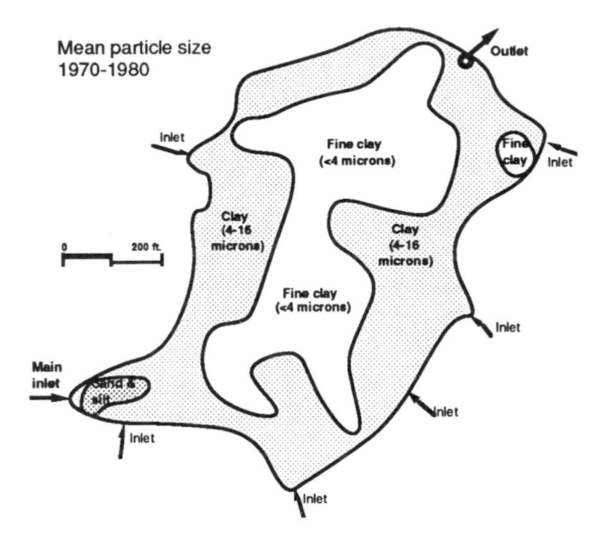

A case study from the Chicago area illustrates how sediment has accumulated in one permanent pool. Lake Ellyn is along a tributary of the East Branch Du Page River. Its area is 10.1 acres; its maximum depth is 6.4 feet. A barrier dam separates the area near the main inlet from the rest of the pond. Eighty percent of the watershed is single-family residential, the remainder in a combination of commercial, parks, and other urban uses. Inflowing sediment originated mostly from soil erosion, decomposition of paved surfaces, and traffic-related sources.

From 1970 to 1980 the pond was monitored to find types and rates of sediment accumulation.

Of suspended sediment this pond removed 91 to 95 percent. Sand settled quickly in the main inlet's forebay. Grading away from all the inlets, sediment accumulated over a 10-year period got shallower, and mean particle size smaller. Most sediment accumulated in the form of an organic-rich mud. Sediment near inlets had petroleum-oil coating and odor.

Of trace metals this pond removed 76 to 94 percent. Trace metals associated with the sediment originated mostly in motor-vehicle traffic and the decomposition of pavement and construction materials. The greatest concentrations of metals were in the deep areas, where the finest sediments were deposited.

Data about Lake Ellyn on this page is from Robert G. Striegl, 1987, Suspended Sediment and Metals Removal from Urban Runoff by a Small Lake, *Water Resources Bulletin* vol. 23, no. 6, pages 985-996.

Trap efficiency is a measure of water-quality control. For a given type of constituent, trap efficiency is the ratio of amount retained in a pond to the total amount flowing in. It can be derived from measurements of mass inflow and outflow over a given period of time from,

$$T = 1 - (Lo / Li)$$

where,
T = trap efficiency (no units);
Lo = load of constituent in outflow, kg; and
Li = load of constituent in outflow, kg.

Little is known about all the physical, chemical and biological reactions that control pond water quality. Hopefully future research will produce more insight. We know that the processes include gravity settling, chemical flocculation and biological uptake by microorganisms. Of these processes, settling of suspended particles is the best understood.

Average residence time of a drop of water in a pond is known to be highly associated with settling of suspended particles and thus sediment trap efficiency. Residence time is the time for a particle of water to flow from the inlet to the outlet, that is, the average time a parcel of water spends in the pond's still-water type of environment.

As residence time increases, a greater proportion of the water's constituents is removed. A time as short as 24 hours can result in significant percentage removal of large particles, at least compared with the few minutes or hours typical of on-site flood-control basins. A residence time of about nine days may remove 70 percent or more of total sediment load. A time of 14 days may be sufficient for settling out 90 percent or more of all suspended solids and their attached pollutants, and significant decomposition of dissolved constituents by algae and microorganisms.

The selection of a residence time for design, like selection of a recurrence interval for a flood-control design storm is judgmental. It is currently even more judgmental and controversial than flood control, since we know so little about what the effects will be.

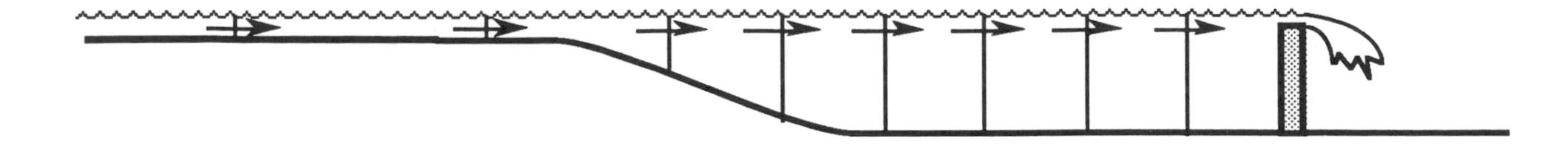

Think of residence time this way:

The volume of water that outlets during a storm is equivalent to the volume that flows in.

But the actual drops of water that flow out are not necessarily the same as the ones that flowed in. In a large pond where the inlet and outlet are far apart, newly inflowing water pushes "old" water ahead of it through the pond and out the outlet. The new water replaces some of the old water in the pond. The volume of the permanent pool remains the same, while the water level gently rises and falls during storms, and individual drops of water within the pond are gradually replaced. Since it takes a number of storms to displace the whole volume, the new water may take a long time to reach the outlet. That time is the water's total residence time in the pond.

Of course a permanent wet pond is necessary for extended detention. Dry ponds are very ineffective at water quality control since the high velocity of flow through them tends to resuspend sediment that may have been deposited earlier, and the residence time for an individual drop of water is much shorter than each individual flow event. In contrast, still water has low velocity to let particles settle out and not to resuspend them.

The volume of the permanent pool must be big enough in relation to the flow volume to reduce velocity and extend residence time. You can imagine the volume of each storm's runoff entering the pool from the upstream end, and displacing an equal volume out the downstream end. The residence time in the pond of each runoff volume would be determined by the size and frequency of the runoff volumes, that is, the average rate of inflow

Average residence time can be controlled through the following relationship:

$$Tr = Vp \,/ Ro$$

where,

Tr = residence time, days;

Vp = volume of permanent pool, ac.ft.;

Ro = inflowing runoff, ac.ft./day

The volume required to create a certain desired residence time is easily derived,

$$Vp = Tr\,Ro$$

The design runoff *Ro* that is used in deriving required volume should be a relatively frequent storm, or a long-term average.

If you use an individual storm event, the storm runoff volume *Qvol* is substituted for *Ro* in the equation for pool volume. Use a relatively small, frequent storm, such as a two-year storm, to assure good quality most of the time. The mean storm (the average storm of all storms that have occurred at a weather station) is used in some areas, such as 0.4 inches in the Washington, D.C. area. A first flush of 0.5 inch has also been used. Using extremely small events does not assure good water quality a very large proportion of the time, nor does it not take into account the possibility of two or more equivalent storms occurring in a short period of time.

Using total monthly runoff is a way to take account of the possibility of a number of equivalent storms occurring in succession. Whatever storm events occur, they are all added up in the monthly total. The largest runoff in ac.ft./day for any month is the design flow. If the pond works during the month with the most runoff, then it will work for the other months as well. The capacity of the pond would be exceeded only in the relatively rare periods with throughflow greater than the largest average month.

A combined approach is to estimate required volume separately for monthly runoff and a selected storm event. The larger of the two required volumes is taken as the minimum design volume.

The adequacy of runoff water supply should be investigated, using the procedures described in the chapter on *Water Harvesting*, to determine whether the required size of pond is supportable. If not, the pond might be lined to reduce infiltration losses, or excavated to groundwater, where present.

A pond's "live" storage area is the area between the pond's outlet and the main inflow point. Throughflow passes through this part of the pond; this is where the replacement of "old" water with "new" takes place. It is often taken to be a rectangle with the length (the distance between inflow and outlet) twice the width. Areas outside this rectangle are "dead" storage in the sense that most inflow will not pass through these areas before it finds its way to the outlet.

In allocating active storage volume *Vs* to a pond, only the live area should be counted. A pond's length:width ratio can be altered to achieve 2:1 by relocating inlets and outlets, creating peninsulas or islands to increase length of flow line, or simply stretching out the pond's length. Some designers have shaped ponds into 2:1 rectangles in order to get 100 percent "live" storage, but this expensive and aesthetically crude approach to "efficiency" is not necessary as long as the pond as a whole contains a live storage area at least as big as the required *Vs*.

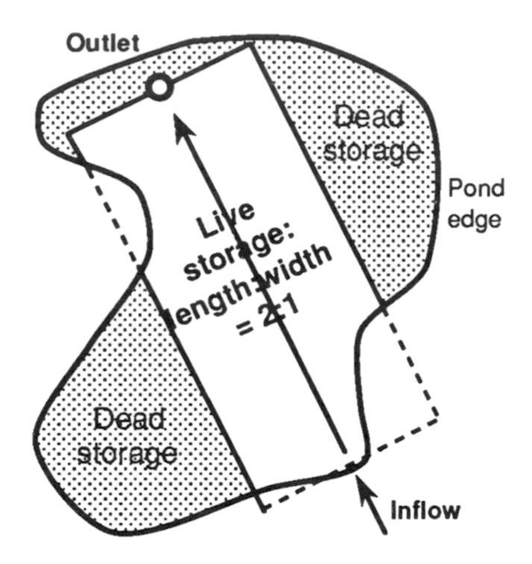

Additional design criteria are sometimes applied within the "live" storage area, to make sure a given pond volume improves water quality effectively.

A large pool area exposes a given volume of water to air and light, giving it opportunity for chemical and biological activity. Some designers have used a certain ratio of pond area to watershed area as a design criterion.

Zones of deep water may give time for settling particles to come into contact with each other and form larger, heavier particles which settle still faster and more permanently.

Zones of shallow water (less than a few feet deep) allow rooting of aquatic vegetation. Here microbiological activity can degrade dissolved pollutants as long as residence time is sufficient. Thirty percent of the pond area in such "littoral" zone has been used as a design criterion is Florida, but this has been based only in imitation of natural lakes, not a knowledge of resulting water-quality effects.

Aeration might help assure decomposition of organic chemicals. In shallow ponds the bottom sediment is naturally aerated. However, in a wide pond that is too shallow bottom sediment tends to be resuspended by wind turbulence.

An early estimate of a pool's area can be a useful step in evaluating extended detention's effect on land use allocation in a development site. Without site-specific topographic information, a pool volume can be assumed to have roughly the shape of an inverted pyramid. If there is no "dead" storage in the pond, its proportions will be 2:1. If we assume that the side slopes in the basin are typical of natural slopes in the region's topography, then we have enough information to derive the basin's top area *At* from the volume *Vp*. The chart below is derived from the geometry of a pyramid. Enter the chart at the bottom with your pond volume, go up to the line for your expected side slopes, thence to the left to read the approximate pond area.

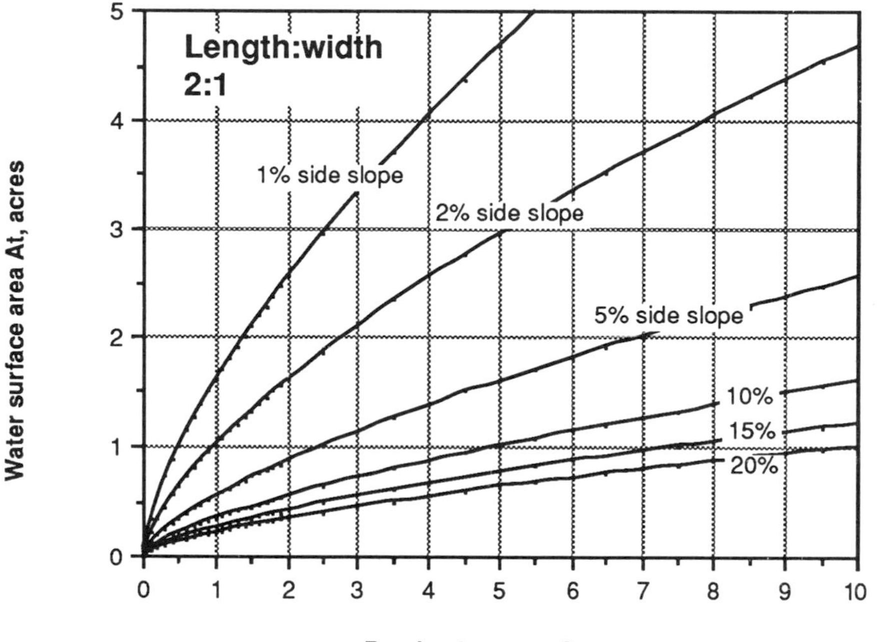

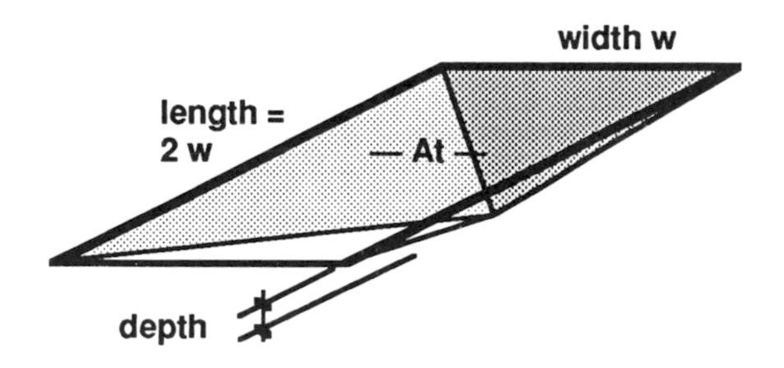

Wetland plants such as bulrush (*Scirpus* sp., *Juncus* sp.), reed (*Phragmites* sp.), cattail (*Typha* sp.) and submerged aquatics can be planted in shallow areas. Dense-growing species provide lots of surface area for microbial life and for uniform distribution of water. Mixed stands of vegetation may remove overall metals most efficiently. However, tall emergent species have been controversial because they may, over their life cycle, deprive pond water of oxygen by contributing only dead decaying vegetative matter to the pond bottom. Where necessary, harvesting aquatic plants every few years may be a way to slow down accumulation of pollutants in a basin.

Natural wetlands can improve the quality of water that passes through them, under certain circumstances. They act as natural extended-detention basins. However, in order not to damage the stability and value of existing wetlands, small quantities of runoff should be discharged into them per area of wetland; they should be protected from overloads of sediment during construction; and stormwater should enter at a number of points via sheet-flow and allowed to distribute throughout the area of the wetland.

Sediment generated during construction can be trapped in an artificial extended-detention pond. The volume of sediment expected to accumulate during construction should be added to the required volume of the pond. In some areas it is a rule of thumb to expect 67 cubic yards of sediment for every acre of watershed, during the entire construction period.

A single basin can fulfill both water quality and flood control functions, if explicitly designed for both. The extended-detention volume is *below* the basin's outlet invert. *Above* the outlet is a volume of "air" held in reserve to control the peak flow of the design storm when it occurs.

The outlet sets the elevation (and hence volume) of the permanent pool, and controls the rate of storm outflow. Like any conveyance, its design should provide for overflow of storms larger than the design storm, and control of erosion at the downstream end. The size of the outlet does not modify the long-term average residence time.

Summary of Process
Extended detention

Define detention objective:

1. Obtain required residence time *Tr* in days from local standards.

Pond volume for monthly runoff:

2. Obtain largest monthly runoff *Ro* in ac.ft./day from your water-balance estimate on page 68, 69, 71 or 72.

3. Compute minimum required pool volume *Vp* in ac.ft. for the wettest month:

$$Vp = Tr\ Ro$$

Pond volume for storm runoff:

4. Obtain 2-year, 24 hour rainfall in inches from page 35.
5. Obtain hydrologic soil group from page 37.
6. Obtain Curve Number from page 39 or 40.
7. Read runoff depth *Dr* in inches from the graph on page 42.
8. Estimate drainage area *Ad* in acres from a site map.
9. Compute storm runoff volume *Qvol* in ac.ft.,

$$Qvol = Dr\ Ad / 12$$

10. Compute required pool volume *Vp* in ac.ft. for the two-year storm event,

$$Vp = Tr\ Ro$$

Design volume and shape:

11. Take the larger of the required monthly or storm-event volumes as the minimum required pool volume.

12. Check that a pond of the required size can be supported by runoff, using the equations described in the chapter on *Water Harvesting*.

13. If applicable, check the surface area of the pond for adequacy of flood control, referring to your storm detention estimates on page 118.

14. Shape the pond on the site such that the required volume fits within the "live" storage area of a 2:1 rectangle.

15. Shape the pond for depth, total area, littoral area and other criteria according to local standards.

Extended Detention Exercise

	Site 1	Site 2
Define detention objective:		
Required residence time *Tr* =	days	days
Pond volume for monthly runoff:		
Largest monthly runoff *Ro* (page 68, 69, 71 or 72) =	ac.ft./day	ac.ft./day
Min. required pool volume *Vp* = *Tr Ro* =	ac.ft.	ac.ft.
Pond volume for storm runoff:		
2-year, 24 hour rainfall (from page 35) =	in	in
Hydrologic soil group *HSG* (from page 37) =		
Curve number *CN* (from page 39 or 40) =		
Runoff depth *Dr* (from page 42) =	in.	in.
Drainage area *Ad* (from site map) =	ac	ac
Runoff volume *Qvol* = *Dr Ad* / 12 =	ac.ft.	ac.ft.
Min. required pool volume *Vp* = *Tr Qvol* =	ac.ft.	ac.ft.
Design pool volume:		
Design volume = larger of monthly *Vp* & storm *Vp* =	ac.ft.	ac.ft.
Assumed side slope in pond (typical of local area) =	%	%
Approximate pool area (from page 127) =	ac.	ac.

Detention Exercises

Discussion of results

1. Which criterion required the larger basin on each site, total monthly run-off or the two-year storm? What general characteristics of the local climate does this point out?

2. Which site requires the larger detention basin? What factors contributed to the requirement of such a large basin?

3. What is the approximate proportion of each site that must be occupied by its basin (basin area ÷ total area)? In your judgment does either basin occupy *too much land* to be reasonable in terms purely of land use?

4. If your response to question number 3 was yes, some approaches you might take to solving the problem include choosing another site for this type of development, modifying the quantitative standard for residence time, and going to some other qualitative type of hydrologic function. Which approach would you most likely choose? What specific site features would you have to build in order to implement your approach?

5. Refer to your storm detention calculations on page 118. Those calculations require a certain pond area in order to accommodate the volume and depth of flood storage. How different is that area from the approximate area of your extended-detention pool? Integrate flood control and water quality control into a single basin by revising the dimensions of one or both of the flood-control and extended-detention basins. What specific sequence of steps do you have to follow to assure that each function will work properly? What size and shape of pond do you end up with? Is additional earthwork or any other construction process necessary to create your integrated pond?

6. Roughly estimate average residence time of storm water in the flood storage volume of a dry storm detention basin, the calculations for which are on page 118. Take into account such factors as the duration of the storm, the average outflow rate indicated by the peak *Qp* out on page 118 or the outflow hydrograph on a computer printout, and the fact that the entire flood volume empties from the basin and flows downstream soon after the rainfall is over. How does this residence time compare with the time you used as a water-quality control criterion? How much improvement in water quality do you expect from a dry storm detention basin? Why?

Summary and Commentary

Extended detention is another scenario for how water could move through a development site, an alternative to conveyance and storm detention. To choose it is to choose a type of hydrologic function, a qualitative way for water to flow through a site.

Extended detention controls quality of surface water. It does *not* affect flow volume (*Qvol*). The entire volume eventually outlets from a basin and continues to be conveyed downstream. Extended detention does not significantly address base flows or groundwater. Downstream from an extended-detention basin the hydrologic function returns to conveyance.

An extended-detention basin is a new facility in a site development program. It must be given adequate space in a site plan, and a positive landscape design.

The results of hydrologic calculations for extended detention give the required volume of basin, and constraints upon length and width. These few results are parameters for site design. As long as you meet these few criteria you are free to *design the site* any way that is needed in terms of materials, planting, contouring, etc.

Chapter 7

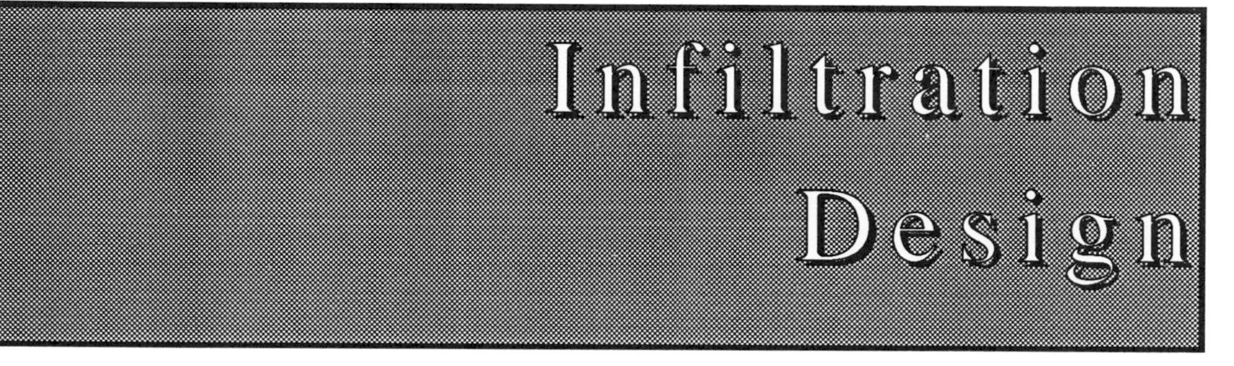

Infiltration is soaking water *into the ground.* It involves capturing storm water in closed basins where it is stored while filtering into the surrounding soil.

The fundamental interest in infiltration comes from a broad concept of water resources. In addition to the flooding and erosion which are addressed by conveyance and detention systems, infiltration can help to control and support groundwater recharge, stream base flows, water quality, and the benefits of aquatic life, recreation, aesthetics and water supplies.

Conveyance and detention maintain *surface types* of flows even though flow *rate* and *quality* are controlled. Infiltration is qualitatively different from conveyance and detention because it puts water into a new *kind of place*, where it undergoes new kinds of processes.

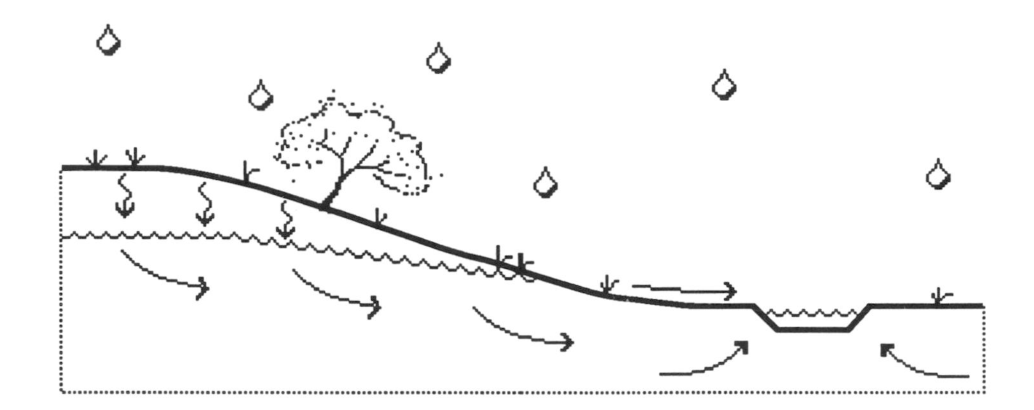

Under natural conditions a large portion of the rain tends to infiltrate the soil and join the soil moisture or, where present, groundwater. In densely vegetated humid regions most rainwater infiltrates where it falls because plants, leaf litter and soil organic matter break up raindrops' energy and let water soak into numerous soil voids. In sparsely vegetated arid regions natural infiltration also operates, except that rain water sometimes has to travel several miles down stream channels before it has time to infiltrate.

Once underground water enters huge "reservoirs" of soil and rock voids. Flow here is very slow compared with that on the surface. Storage here is typically very large compared to anything available on the surface.

Water may reside in the subsurface for weeks or months before reaching streams. Water emerges from subsurface storage so gradually that its discharge is part of the streams' base flow, not the original storm flow. The flow is *qualitatively* different. It has become a different *type* of flow, supporting different types of resources.

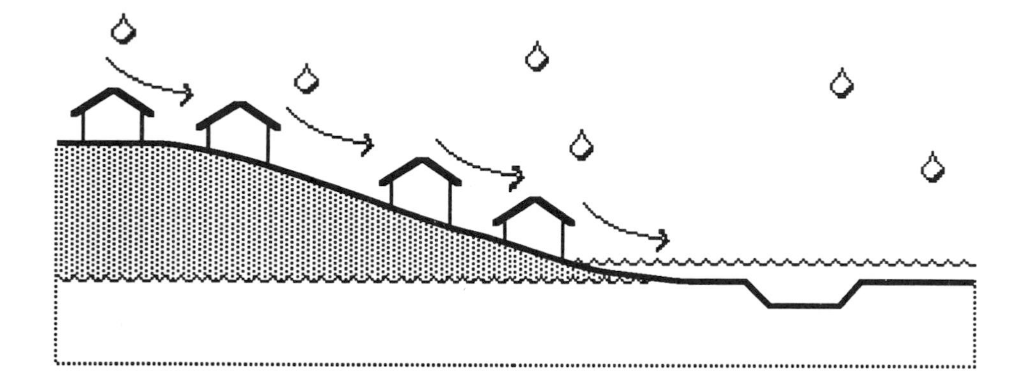

When land is developed new impervious surfaces deflect rainfall from its natural course.

Large quantities of surface runoff are sent directly to streams during storm events, making flood flows larger. Runoff water moves quickly down stream systems and out of the regions where it originated.

Little subsurface storage can occur. Rain water that used to infiltrate is deflected away. Groundwater levels, soil moisture and stream base flows decline.

Thus development both aggravates the *hazards* associated with storm flows and injures the *resources* supported by low flows. Fluctuation in stream flow is as natural as the falling of leaves in the autumn. But development aggravates natural fluctuation, making highs higher and lows lower. Development prevents water from entering the natural processes of filtration, storage and gradual discharge that moderate hydrologic flows.

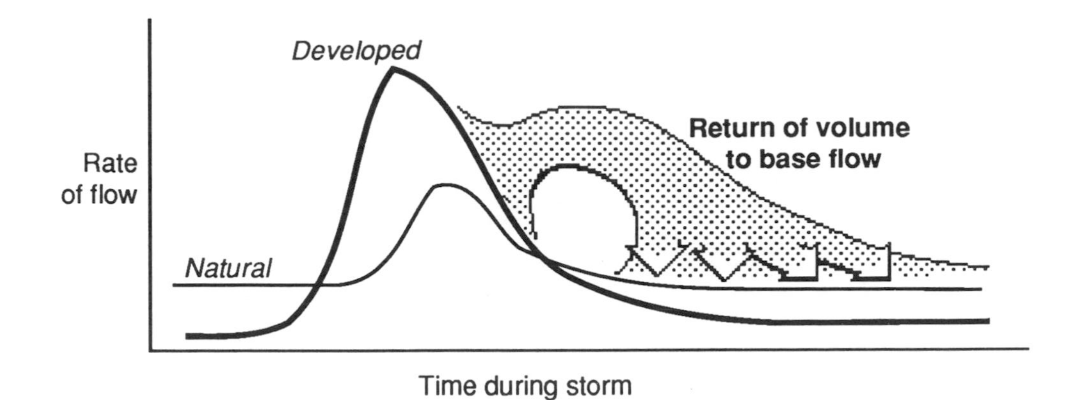

With this background one of the mechanisms by which development changes storm hydrographs can be understood. After houses and roads are constructed the *rainfall* on a site is the same as it was before. But *more water* must have come from somewhere, because both the peak rate of flow and the total volume of stormflow are higher. Where did the extra water come from?

The "extra" water was there on the site all along. Under natural conditions it infiltrated the soil and emerged long after storms were over. Impervious surfaces deflect that water into storm flow. Piling up all that volume in a compressed time necessarily creates a higher peak flow rate. The storm hydrograph, which everyone has known for a long time shows the addition in the peak rate of flow *during the storm event,* also shows the volume of flow that has been diverted *away* from low-flow periods *between* storms.

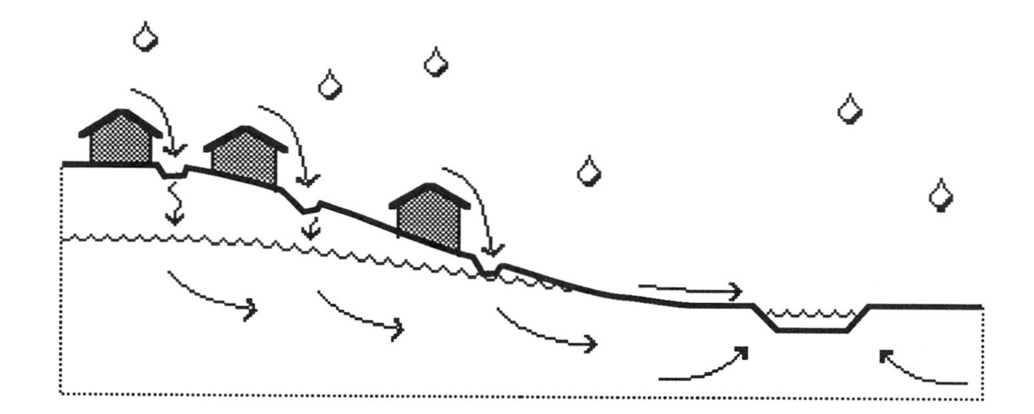

So if we are concerned about a broad range of water resources and not just floods, then we have to be concerned about the *volume* of storm flow that goes into streams as well as the *peak rates.*

Stormwater infiltration is aimed at *volume* of runoff so that it can keep aggravated storm flows out of streams and return them to their native place in the soil and long-term base flows. Reducing volume of flow also reduces peak flow rate.

Infiltration can reclaim storm runoff. By forcing runoff into the soil, it returns rain water to the path it took before development occurred. In some cases infiltration may recharge *more* groundwater than before development, by forcing into recharge some of the water that would have been lost to evapotranspiration. Infiltration returns runoff to its natural place in soil moisture and groundwater, and takes advantage of the land's natural capacities for filtration and long-term storage. It uses the subsurface environment as a resource to reduce flood flows and to sustain and purify low flows in streams and groundwater.

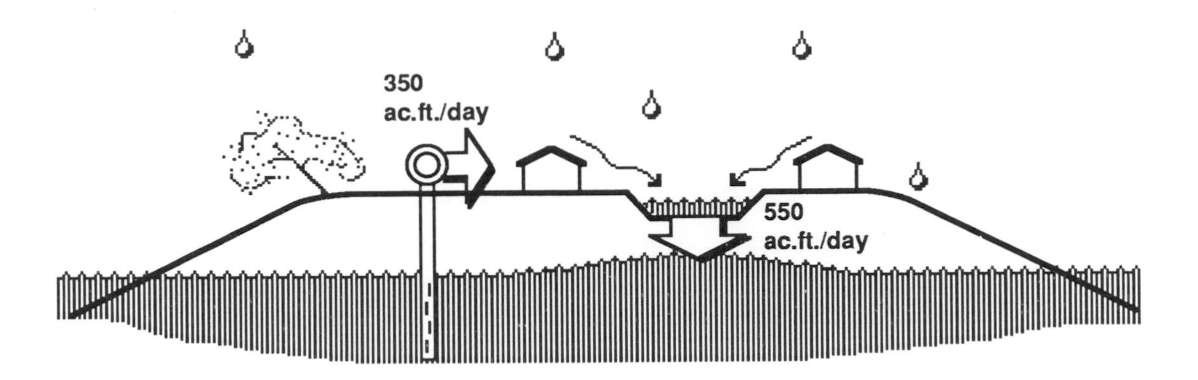

Long Island has been infiltrating urban stormwater since the 1930s. Its original motive for "recharge" basins was to avoid expensive storm sewers by eliminating runoff near the source. But since then residents have found that the basins help maintain the Island's groundwater, which is their only water source.

The hydrologic effect of the more than 3,000 basins that were in place as of 1980 can be shown by a simple analysis. The areas that the basins drain add up to about 110,000 acres. This is more that 10 percent of the area of the counties (Nassau and Suffolk) where they are used. The development in the basins' drainage areas averages about 2.5 dwelling units per acre, housing a total population of about 750,000 residents. At typical rates of water use in the region, a groundwater withdrawal of about 350 acre-feet per day can be attributed to the persons who live in the areas. The land cover in the drainage areas is about 25% impervious. Of the 45 inches of rain per year that falls on the drainage areas, about half, or 22 inches, infiltrates through the recharge basins. This equals or exceed predevelopment recharge, and amounts to a groundwater replacement equal to 550 acre-feet per day.

Long Island's storm water is being recycled. Recharge basins replace ground water about 50 percent *faster* than the population of the basins' drainage areas withdraw it. As long as future developments include similar basins, urban areas can continue to grow without exceeding the Island's native water supply.

Figures about Long Island on this page were derived from data in G.E. Seaburn and D.A. Aronson, 1974, *Influence of Recharge Basins on the Hydrology of Nassau and Suffolk Counties, Long Island, N.Y.*, U.S. Geological Survey Water-Supply Paper 2031; Wayne B. Solley and others, 1983, *Estimated Use of Water in the United States in 1980*, U.S. Geological Survey Circular 1001; and Henry F. Ku and Dale L. Simmons, 1986, *Effect of Urban Stormwater Runoff on Ground Water Beneath Recharge Basins on Long Island, New York*, U.S. Geological Survey Water-Resources Investigation Report 85-4088.

Water quality effects of infiltration have also been documented on Long Island and in other parts of the country. Infiltration treats runoff *before* it reaches streams or groundwater.

Long Island's underlying aquifer has not been measurably polluted by stormwater infiltration basins, either chemically or microbiologically. In fact, some basins have been found to *dilute* nitrogen in the groundwater with fresh water, where the aquifer was already polluted to a degree with nitrogen from other sources such as septic tanks.

In the Fresno area of California's Central Valley similar results were found. Basins up to about 20 years old were found not to be associated with significant contamination of soil water percolating through the valley's alluvial soils. Nor were they associated with pollution of the underlying groundwater. Concentration of constituents in the groundwater under the basins was similar to levels reported in the regional groundwater.

In the upper few inches of a basin's soil metals, phosphorus and other constituents tend to accumulate, mostly in association with the soil's clay portion. Nitrogen tends to decompose and return to the atmosphere. In watersheds with ordinary urban pollutant loadings this accumulation implies that sediment ought to be removed from basin floors after about 10 to 50 years, although no agency has yet found it actually necessary to undertake such an expense with existing basins.

Aeration can help assure decomposition of pollutants. Where a basin is above the water table, a clearance of a few feet over the water table leaves an unsaturated zone of soil where aerobic decomposition can occur. In Florida, "retention" basins are regularly excavated into the water table, making them appear as permanent ponds; decomposition of pollutants while still in the ponds is adequate when the ponds have zones shallow enough (less than a few feet) that the bottom sediment is naturally aerated.

Large shopping-center parking lots generate extraordinarily high pollutant loads. In some of Long Island's basins with such watersheds, the surface soil has become clogged by adhering petroleum-based matter. Long Island's response in these cases has been to retrofit the drainage system by using the clogged basins only as settling basins, abandoning their use for infiltration, and letting the clean overflow run into new, adjacent infiltration basins. Presumably it would be possible to plan a similar approach, involving a settling basin upstream of an infiltration basin, for proposed drainage systems as well.

Data about Long Island on this page are from Henry F. Ku and Dale L. Simmons, 1986, *Effect of Urban Stormwater Runoff on Ground Water Beneath Recharge Basins on Long Island, New York*, U.S. Geological Survey Water-Resources Investigations Report 85-4088. Data about Fresno, California are from Harry I. Nightingale, 1987, Water Quality Beneath Urban Runoff Management Basins, *Water Resources Bulletin* vol. 23, no. 2, pages 197-205; Harry I. Nightingale, 1987, Accumulation of As, Ni, Cu, and Pb in Retention and Recharge Basins from Urban Runoff, *Water Resources Bulletin* vol. 23, no. 4, pages 663-672; and Fresno Metropolitan Flood Control District, 1977, *Infiltration Drainage Design for Highway Facilities*, Fresno: Fresno Metropolitan Flood Control District. Data about Florida are from Yousef A. Yousef and others, 1986, Nutrient Transformations in Retention/Detention Ponds Receiving Highway Runoff, *Journal Water Pollution Control Federation* vol. 58, no. 8, pages 838-844; and Yousef A. Yousef and others, Fate of Heavy Metals in Stormwater Runoff from Highway Bridges, *Science of the Total Environment* vol. 33, pages 233-244.

A basin excavated in the earth holds the runoff that reaches it until it soaks in. Unlike a detention basin, an infiltration basin has no regular surface outlet. The continuity of the invert along the primary drainage system is broken. Site grading is broken into discrete drainage areas that flow into each other only when the design storm is exceeded. When numerous basins eliminate runoff near its source, and each basin is sized to hold a design storm as well as all background flows, the expense of culverts and swales to convey runoff downstream is eliminated.

An open dry basin can be used in a development where there is sufficient open space, and a level area of earth or grass would be seen as a multiple-purpose amenity. A large level area is necessary to eliminate standing water quickly after storms by infiltration and evaporation. It is immediately accessible for inspection and maintenance. Vegetating it with grass or other plants tolerant of occasional flooding, such as Bermuda grass, can help to maintain soil porosity. It can be multiple-use if properly located and shaped. Although the construction of an open dry basin involves few costs other than proper grading and ordinary planting, its large area may preempt the economic use of valuable urban land.

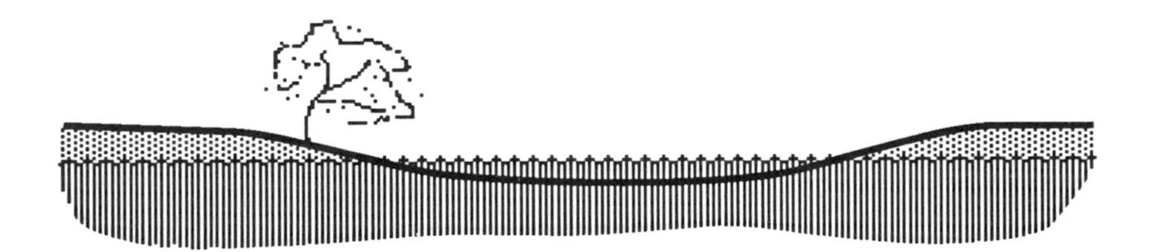

A permanent pool with a relatively stable water level can be formed where an infiltration basin's floor intersects the groundwater table. Stormwater can be temporarily mounded on top of the water table until it infiltrates through the basin sides and by pushing down and spreading out the underlying water table. The temporary storage volume above the seasonal high water table must be capable of receiving the volume of the design storm. A permanent pool can enhance community amenities if thoughtfully located and shaped.

On Longboat Key, Florida, the Buttonwood Cove residential development left an old lowland grove of cypress trees intact, only the ground cover being modified with some grass and gravel. Some of the development's runoff drains into the natural low area and infiltrates without disrupting the ornamental surface covers or the native canopy trees.

Near Syracuse, New York, the Anaren Microwave industrial headquarters is on a glacial lake plain with shallow groundwater and almost no topographic relief. Swampy areas were dredged to generate fill for raising the building elevation, and the borrow areas shaped to form open, tree-lined ponds. The ponds' water level fluctuates seasonally with the groundwater table. All runoff is drained into the ponds, where it mounds up until it dissipates into the surrounding groundwater. There is no surface outlet from the site.

In West Palm Beach, Florida, a landfill is ringed with excavated "retention" basins to prevent potentially contaminated runoff from directly entering nearby wetlands. Groundwater makes the basins into permanent pools. To mitigate wetland destruction inevitably resulting from construction of the landfill, thousands of cypresses and other wetland plants were transplanted from the landfill site to the shores of the retention basins, where they compose an artificial wetland accessible to the public and to wildlife.

The quantitative sizing of such dry and wet basins is described in the chapter on *Water Harvesting*, because of their multiple-use intent.

An ephemeral basin has a fluctuating water level, or only occasional standing water.

A hydraulic advantage of a basin designed for an ephemeral regime is infiltration through basin walls as well as the floor. This provides a buffer against low infiltration rates on basin floors which may result from compaction and sediment.

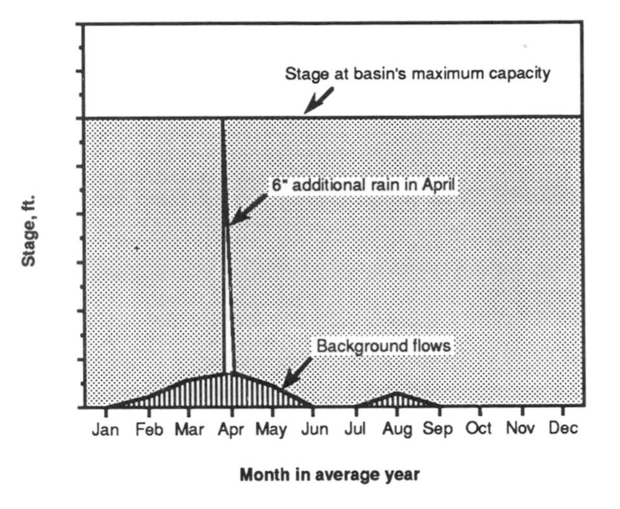

The graph shows ephemeral occurrence of water in one basin in the eastern United States. The darker tone represents the volume of water stored in the basin in an average year; a seasonal pattern is clear. The lighter tone represents unused capacity held in reserve for a design storm. If in April, when the water level in this basin is highest, a 10-year design storm occurs in addition to the average monthly precipitation, the basin is filled to its capacity. Due to infiltration from the basin, full reserve capacity becomes available within a month after the storm.

Open ephemeral basins in wetland-rich Florida are interpreted as part of the "natural" environment, and the water birds that frequent them are much admired. In other regions of the country the mud, mosquitoes and safety hazards perceived to be associated with temporary pools may not be tolerated by urban residents.

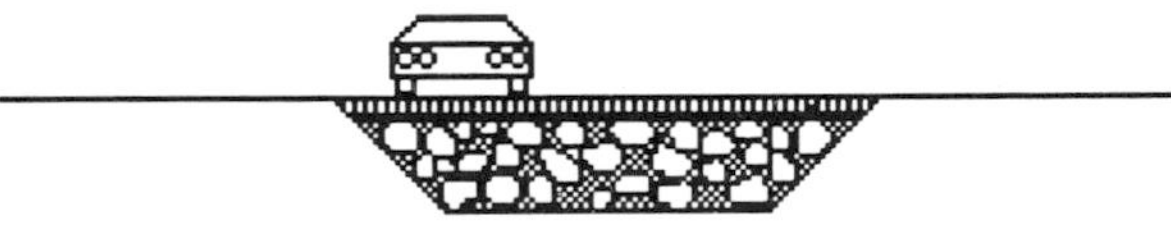

Underground basins can isolate perceived nuisances from the surface environment. Where little open space is available, underground basins allow the surface to be reclaimed for economic or aesthetic uses. They infiltrate a large proportion of annual inflows, since they protect collected water from evapotranspiration.

Locate underground basins under level paved areas such as parking lots or sand-covered playgrounds. Specifically, locations under the parking *spaces* of parking lots have relatively low traffic loads. Locate basins away from building foundations, leaching fields and steep cut or fill slopes. In areas of swelling soil, a surface of loose aggregate over a basin can take up movements without functional or aesthetic damage.

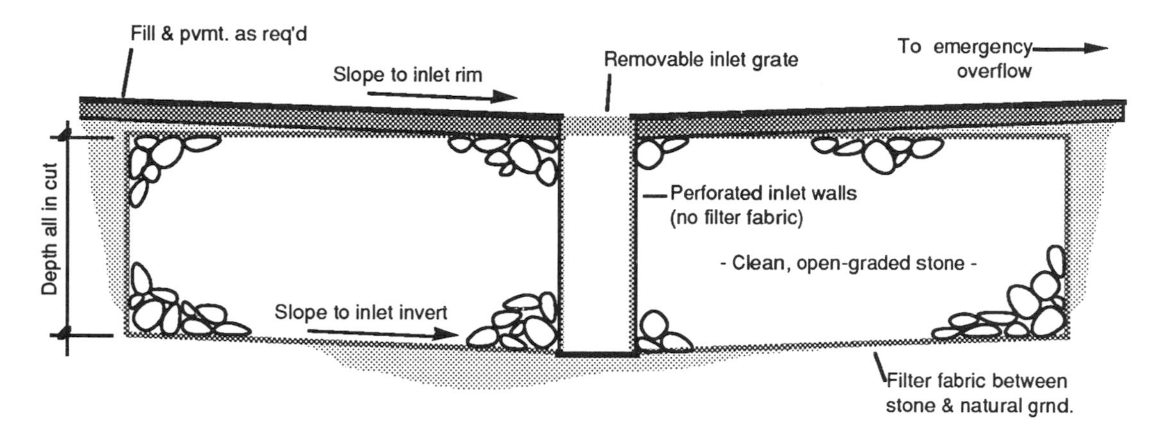

An underground basin can be backfilled with stone to support surface loads. Stormwater fills the stone's void spaces until it infiltrates the surrounding soil. The stone fill must be clean, open-graded crushed stone, such as No. 4 or No. 5 stone. Wrapping the stone in an envelope of drainage filter fabric prevents surrounding soil from collapsing into stone voids.

The volume of a stone-filled basin must equal the sum of the volumes of water and solid rock. Open-graded crushed stone has 40 percent void space, so the total volume of the basin must be 2.5 times the volume of water it is meant to hold.

Use drainage inlets to let in runoff. Perforated inlet walls admit water freely into the stone fill. Removable inlet grates allow inspection and maintenance access. Slope the floor downward to the bottom of each inlet to make sediment and standing water accumulate only where they can be accessed for maintenance.

Avoid passing concentrations of raw stormwater though a functional or ornamental surface such as a "porous pavement" or a decorative stone or sand mulch. Sediment in the water will constantly attempt to stain and disfigure the decorative surface and clog its pores. Seeds carried in the sediment will sprout in pockets of moist silt. Moving traffic will constantly "work" an asphalt-based pavement, filling any pore spaces.

In Glen Burnie, Maryland, the Cromwell Field shopping center has a parking lot built over a stone-filled basin six feet deep and an acre in area. All that is visible from the surface is conventional drainage inlets. The surface on this valuable, intensely developed site was reclaimed for economic use.

Near Orlando, Florida, the Sunbay Club multifamily residential development has tennis courts underlain by a gallery of parallel perforated pipes. Runoff is let into the pipes from the surrounding stormwater collection system. The pipes are surrounded by stone to enlarge the infiltration surface area. Infiltrating water recharges the valuable Floridan aquifer. Runoff passes through a series of ornamental fountains and pools before reaching the infiltration gallery, which probably helps aerate the water and remove sediment.

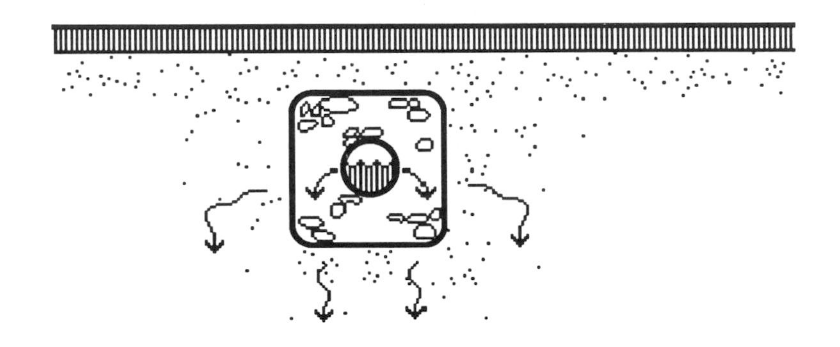

The sizing of such underground, stone-filled ephemeral basins is covered on the following pages. Their routing is relatively simple, because the overlying pavement or other material makes evaporation and direct precipitation negligible. For open, exposed basins, where these additional flows are significant, see the appendix on *Open Ephemeral Basins.*

The design storm must be contained within the volume of an infiltration basin. This controls floods without a separate detention basin, and eliminates the need for a primary conveyance system outside the basin. Emergency overflow must be provided for in an infiltration basin as in any other drainage facility.

Low-level background flows must also be contained in an infiltration basin. An infiltration basin accumulates all flows, large and small. Capturing the low-level everyday flows controls water quality and assures continuous subsurface recharge. Taking into account the volume of background flows assures that a basin's capacity is not preempted by standing water when the design storm occurs.

An ephemeral basin's monthly changes in stage and storage can be estimated by applying the water balance. In principle a monthly infiltration routing is like any other reservoir routing, in that it applies the fundamental reservoir equation,

$$\Delta\text{Storage} = \text{Inflow} - \text{Outflow}$$

This approach originated in Minnesota about 1980. Minnesota's Department of Transportation wanted to find out whether highway runoff could be disposed in existing glacial kettle holes. They used a long-term average water balance to estimate monthly water levels. The 100-year storm volume was then superimposed on the highest monthly water level. If a basin did not overflow under these conditions, it was considered capable of disposing the runoff.

Proposed basins can be sized using the same type of procedure. Trial basin dimensions are chosen, such that the basin's capacity is at least equal to the volume of the design storm and the volume of stone fill has been taken into account. The routing begins in the month with the greatest water deficit (excess of evaporation over precipitation), with initial stage and storage set at zero. Runoff inflows and evaporation outflows create a change in storage and stage. The stage at the end of each month is carried to the beginning of the following month. Each new monthly stage determines how much side area is in contact with water, and thus the soil area for water to infiltrate into. (The infiltration rate through the floor may be lower than that through the sides because of compaction and sediment accumulation; this can be taken into account by applying a safety factor of about 0.5 to the floor infiltration rate.) To correct any possible error in the assumed initial stage and storage, the cycle of average months is repeated until each month's stage is the same it was in the same month the previous year. At this point the basin's hydrologic regime is adapted to long-term average conditions.

A trial design is evaluated by comparing the monthly unfilled volume to the reserve capacity necessary to hold the volume of the design storm. If necessary the basin dimensions are adjusted, and the water balance is recalculated, until a satisfactory basin is found.

The Minnesota method was first described in Daniel Engstrand, 1983, ***Retention Ponds: Analysis and Design of Ponds without Outlets***, St. Paul: Minnesota Dept. of Transportation, and developed in Bruce K. Ferguson, 1989 (in press), Role of Long-Term Water Balance in Management of Stormwater Infiltration, *Journal of Environmental Management*.

Pure design-storm approaches to infiltration have been tried in some areas. However, such approaches disregard the low-level background flows that occur between the major storms.

"First-flush" approaches are an attempt to control water quality. Infiltration basins with small capacity, such as a 1/2 inch runoff volume, are intended to capture the most highly polluted water. However, background flows could preempt part or all of the capacity many months during the year, causing all the remaining runoff, both first-flush and otherwise, to overflow and continue down the stream system.

Maryland in 1983 required that developers infiltrate the increase in design-storm runoff volume attributable to their developments. Three years later, basins that had been sized for that standard were inspected on random days. A third of them were found to contain inadvertent standing water, which reduced their capacities and, in above-ground basins, created nuisances. Low-level background flows, which had not been taken into account in design, had preempted the capacities of the basins.

Any of the infiltration objectives found in various regions of the country are technically feasible as long as they are supplemented by the long-term water balance. Capture and infiltration of the highly polluted first flush is possible by sizing a basin to contain all monthly background flows. Capture of a design storm is possible by sizing a basin for both the accumulated background flows and the storm flow itself.

Data on this page are from Bruce K. Ferguson, 1989, Role of Long-Term Water Balance in Management of Stormwater Infiltration, *Journal of Environmental Management.*, and L. Kenneth Pensyl, and Paul F. Clement, 1987, ***Results of the State of Maryland Infiltration Practices Survey***, Annapolis: Maryland Department of Natural Resources, Water Resources Administration, Stormwater Management Division.

Subsurface features such as bedrock or impermeable soil layers may limit the way you can infiltrate on a specific site. Where a soil survey suggests that such a condition is present you can keep the *bottom* of basins at least a few feet *above* the limiting layer to permit positive infiltration.

Shallow impermeable layers that inhibit infiltration sometimes overlie permeable material into which you want the water to infiltrate. Some basins on Long Island have addressed this by puncturing the impermeable layer, and providing a conduit for water to pass into the underlying aquifer.

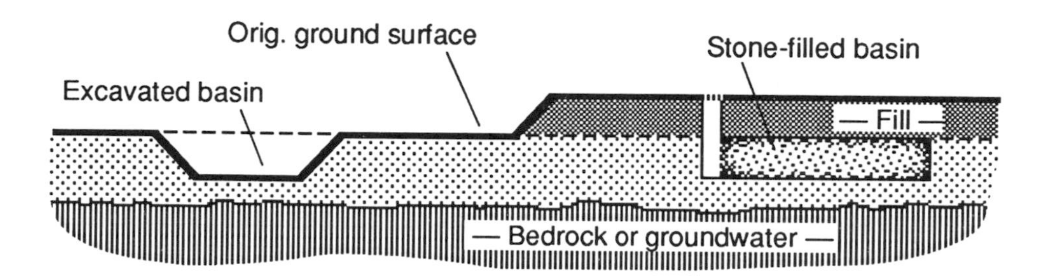

Where grading has excavated below the original surface to make room for roads or buildings, the depth of cut must be subtracted from the permissible depth of basin below the original surface.

Where grading has filled you should avoid infiltrating into structural fills such as those that support roads or buildings. However you can excavate a basin through a noncritical fill to infiltrate into the original soil.

Summary of Process:
Underground, stone-filled infiltration basin

Trial basin:

1. Obtain design-storm runoff volume (*Qvol*) in ac.ft. after development, from storm runoff estimate on page 50.

2. Compute basin volume for design storm including stone fill from,

$$\text{Total basin volume} = Qvol/0.4$$

3. Set initial trial basin dimensions of a rectangular basin (length *L*, width *W* and depth *Dt*) in feet to make the basin at least as big as the total needed for a design storm.

4. Compute floor area *Af*,

$$Af = \text{Total volume} / Dt$$

5. Compute side area *As* from the geometry of the basin.

Monthly storage:

6. Identify the month with the greatest water deficit (the greatest excess of lake evaporation over precipitation) from the data used in water balance estimates on pages 68, 69, 71 or 72, and 76.

7. In the month with the greatest deficit, set initial basin stage *S*, storage *Vw*, and wet side area *As* wet all equal to zero.

8. Obtain monthly volume of inflowing runoff *Ro* in ac.ft. from your water-balance estimate on page 68, 69, 71 or 72.

9. Compute monthly volume of outflowing infiltration *Ivol*. With infiltration *I* in ft./mo. obtained from your estimate on page 76, a selected floor infiltration safety factor *Fs*, and a different hydraulic gradient *Ghf* from that on the sides *Ghs*, the infiltration volume in ac.ft./mo. is given by,

$$Ivol = (GhsAs\ \text{wet} + FsGhfAf)I$$

10. Compute monthly change in water storage ΔVw,

$$\Delta Vw = Ro - Ivol$$

11. Compute monthly change in *S* in ft. from the void ratio and basin geometry,

$$\Delta\text{Stage} = (\Delta Vw\ /\text{void ratio}) / Af$$

12. Compute the stage *S* in ft. at the beginning of the following month,

$$S = \text{Previous } S + \Delta\text{Stage}$$

13. Compute the wet side area *As* wet in acres at the beginning of the following month,

$$As\ \text{wet} = S\ (L + W) / 21{,}780$$

14. Compute the unfilled volume held in reserve for a design storm in ac.ft.,

$$\text{Reserve capacity} = (Dt - S)\ (Af)\ (\text{void ratio})$$

15. Repeat the routing calculations until the stage at the end of each month is the same as that in the same month of the previous year.

Evaluation:

16. Compare the basin's minimum monthly unfilled volume to the reserve capacity needed to hold the volume of the design storm.

17. If necessary, adjust the basin's dimensions and revise the routing calculations accordingly.

Infiltration Exercise
Underground, stone-filled basin

Site 1

Design-storm volume $Qvol$ (from page 50) = ________ ac.ft.; Trial capacity = ________ ac.ft.

Void ratio = ________; Total volume Vt = Capacity / Void ratio = ________ ac.ft.

Trial dimensions: Total depth Dt = ________ ft.; Floor area $Af = Vt / Dt$ = ________ ac.; Length L = ________ ft.; Width W = ________ ft.

Floor infiltration safety factor Fs = ________; Floor hydraulic gradient Ghf = ________; Side hydraulic gradient Ghs = ________

	Jan	Feb	Mar	Apr	May	Jun	Jul	Aug	Sep	Oct	Nov	Dec	Annual
Conditions at beginning of month:													
Stage S, ft. = previous stage + Δ stage (not less than zero)	___	___	___	___	___	___	___	___	___	___	___	___	
Wet side area As wet, ac. $= S\ (L + W) / 21{,}780$	___	___	___	___	___	___	___	___	___	___	___	___	
Reserve capacity, ac.ft. $= (Dt - S)(Af)$(void ratio)	___	___	___	___	___	___	___	___	___	___	___	___	
Inflows and outflows of water:													
Runoff inflow Ro, ac.ft./mo. (from page 68 or 71)	___	___	___	___	___	___	___	___	___	___	___	___	___
Infiltration I, ft./mo. (from page 76)	___	___	___	___	___	___	___	___	___	___	___	___	
Infiltration volume $Ivol$, ac.ft./mo. $= (GhsAs \text{ wet} + FsGhfAf)\ I$	___	___	___	___	___	___	___	___	___	___	___	___	___
Change in storage ΔVw, ac.ft./mo. $= Ro - Ivol$	___	___	___	___	___	___	___	___	___	___	___	___	
Change in stage:													
Δ stage, ft./mo. $= (\Delta Vw / \text{void ratio}) / Af$	___	___	___	___	___	___	___	___	___	___	___	___	___

Infiltration Exercise

Underground, stone-filled basin

Site 2

Design-storm volume $Qvol$ (from page 50) = ________ ac.ft.; Trial capacity = ________ ac.ft.

Void ratio = ________; Total volume Vt = Capacity / Void ratio = ________ ac.ft.

Trial dimensions: Total depth Dt = ________ ft.; Floor area $Af = Vt / Dt$ = ________ ac.; Length L = ________ ft.; Width W = ________ ft.

Floor infiltration safety factor Fs = ________; Floor hydraulic gradient Ghf = ________; Side hydraulic gradient Ghs = ________

	Jan	Feb	Mar	Apr	May	Jun	Jul	Aug	Sep	Oct	Nov	Dec	Annual
Conditions at beginning of month:													
Stage S, ft. = previous stage + Δ stage (not less than zero)	___	___	___	___	___	___	___	___	___	___	___	___	
Wet side area As wet, ac. $= S\ (L + W) / 21{,}780$	___	___	___	___	___	___	___	___	___	___	___	___	
Reserve capacity, ac.ft. $= (Dt - S)(Af)$(void ratio)	___	___	___	___	___	___	___	___	___	___	___	___	
Inflows and outflows of water:													
Runoff inflow Ro, ac.ft./mo. (from page 69 or 72)	___	___	___	___	___	___	___	___	___	___	___	___	___
Infiltration I, ft./mo. (from page 76)	___	___	___	___	___	___	___	___	___	___	___	___	
Infiltration volume $Ivol$, ac.ft./mo. $= (GhsAs\text{ wet} + FsGhfAf)\ I$	___	___	___	___	___	___	___	___	___	___	___	___	___
Change in storage ΔVw, ac.ft./mo. $= Ro - Ivol$	___	___	___	___	___	___	___	___	___	___	___	___	
Change in stage:													
Δ stage, ft./mo. $= (\Delta Vw$ /void ratio) / Af	___	___	___	___	___	___	___	___	___	___	___	___	___

Infiltration Exercise:
Discussion of Results

1. What proportion of each site would be occupied by its basin (basin area ÷ total area)? In your judgment does either of the basins occupy *too large* a proportion of its site, or require too big a volume of stone, to be reasonable in terms of land use, site design or cost? If so, what conditions contributed to the requirement of such a large basin? What approaches to hydrologic function or site design could you take to resolve this predicament?

2. Draw a graph with months on the *x* (horizontal) axis and monthly stage in your basin as the *y* (vertical) axis. In which months is stage the highest? The lowest? Are there existing basins, reservoirs or lakes in your area that have the same type of regime? What characteristics of the regional climate contribute to such a regime?

3. Write an inspection and maintenance program for your basin. Specify frequency and timing of inspection for standing water and sediment accumulation, and what types of maintenance or repair are to be made when specific conditions are found.

4. Compare the hydrologic cycle on your proposed site with natural sites in the surrounding region, as follows. Find the average annual precipitation, runoff and evapotranspiration in your area: precipitation in inches is in local weather data such as those used in the water balance calculation; runoff in inches can be found from a recent edition of the U.S. Geological Survey's *Water Resources Data* for your state (select from that publication a stream near you, preferably with a watershed area not more than about a hundred square miles so that it is all in the same climatic region, and without major artificial withdrawals); evapotranspiration can then be pretty closely approximated as the difference between precipitation and runoff. Then estimate the amount of water that will infiltrate through your basins by adding up the monthly infiltration volumes *Ivol*; subtract annual infiltration from annual precipitation to get evapotranspiration on your site. What is the difference between evapotranspiration in the region, and evapotranspiration on your site? Since precipitation is presumably the same on your site as in the surrounding region, what happens to any rain water that does not evapotranspire? What characteristics of your watershed and your basin contribute to this contrast with the natural pattern? Does the change in hydrologic cycle represent a problem, or an improvement? In what way?

Summary and Commentary

Infiltration is a new kind of objective for stormwater management. It is qualitatively different from conveyance and detention because it puts stormwater in a different part of the environment.

Infiltration is still a form of disposal because stormwater ends up being discharged to the environment.

However, we can say that it is discharged to a part of the environment where it is more useful and accessible than if it were dumped into a fast-moving stream.

After leaving the site, infiltrated water supports local base flows and groundwater. It does not combine with surface flows to aggravate flooding during the storm period. The soil filters contaminants before they reach groundwater or streams.

Infiltration basins comprise a distinct type of facility in a site development program. Specific spaces and construction costs must be reserved for them in site plans.

Hydraulics need not dictate urban design. The results of infiltration's hydrologic design calculations are basins' length, width and depth. These dimensions must be coordinated with the site design and grading plans of the sites where the basins are located. As long as you meet these few quantitative criteria you are free to *design the site* any way that is needed in terms of materials, contouring, planting, etc.

Chapter 7

Water harvesting is collecting runoff for positive use. When considered alone, it is aimed not at control of negative stormwater impacts but at using runoff for positive water supply. Water harvesting's fundamental objective of on-site collection for use can be supplemented by control of the quantity or quality of stormwater leaving the site, as long as both are taken explicitly into account. The term "water havesting" started with plant irrigation in the arid Southwest; the same term can be used for the supply of water to ponds or any other on-site uses in other parts of the country.

A complete change of viewpoint is required. Instead of assessing *drainage areas* as sources of nuisance runoff we plan *catchment areas* to generate water supplies. Instead of minimizing runoff *impact* we encourage runoff *efficiency* to get a large water supply from the available rainfall. Instead of diverting runoff *away* from the areas where it could do damage we direct it *toward* points of use.

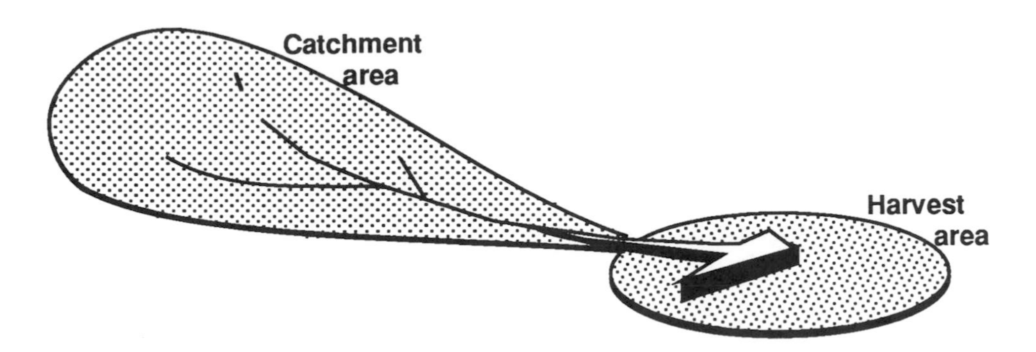

Adequacy of water supply is determined by the quantity of runoff relative to expected losses during use. The balance between supply of inflowing water from the *catchment area* and losses to outflows from the *harvest area* limits the sizes of basins, ponds and irrigated areas that can be supported.

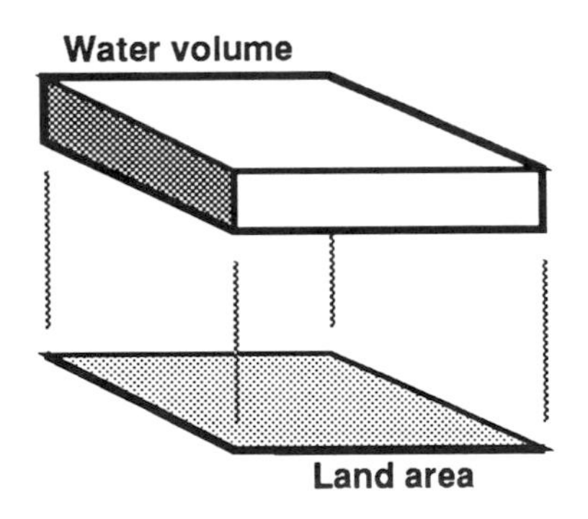

When the volume of loss is proportional to harvest area, it is possible to derive the area that equalizes outflows with inflow in any one individual month.

A design equation to find the size of harvest area with outflows equal to inflows is in the form,

$$\text{Area} = \text{Runoff inflow} / \text{Sum of outflows}$$

The supply of harvested water is monthly runoff *Ro* in ac.ft./mo.; for a given catchment area it is a fixed quantity. Monthly outflows include such losses as evaporation and seepage in units of ft./mo.; their volumes are controlled by sizing the land area where they occur.

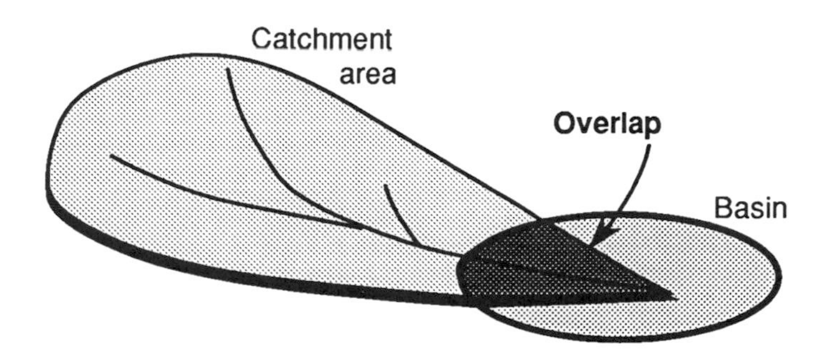

A harvest area might overlap, to a degree, the catchment area that produces its runoff supply. As such an area gets larger, it effectively reduces its own runoff inflow by reducing the area of the catchment. The reduction in volume of water is equal to the overlapping area times the depth of runoff from the catchment. This effect can be taken into account by roughly estimating, from a site plan, the proportion X of a basin that will overlap its catchment area. Loss of runoff depth Ldr due to catchment overlap is then proportional to X:. It is measured in ft., and is counted in a design equation along with other losses. With Ldr in ft., Dr in ft., and X being a decimal fraction with no units,

$$Ldr = Dr\,X$$

Direct precipitation upon a harvest area is an inflow which supplements catchment runoff. However, like the outflows, it is measured in ft. rather than ac.ft.; the volume of water it contributes shrinks and grows with the size of the harvest area. Consequently, in applying a design equation, direct precipitation is counted as a reduction of losses rather than an addition to outflow, by subtracting it from actual losses such as evaporation and infiltration.

Dry Infiltration Basins

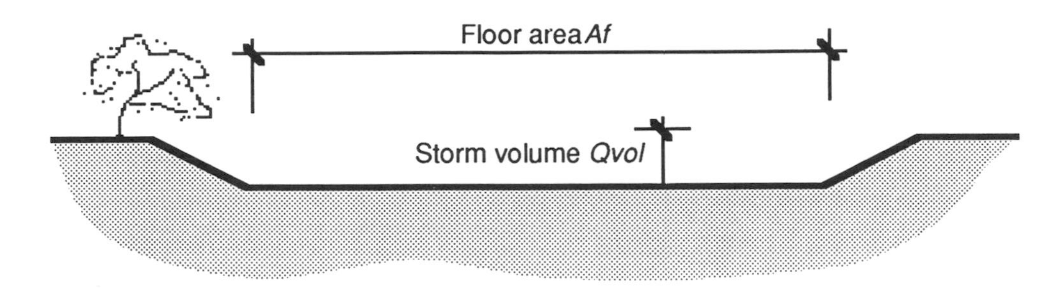

Dry regimes in infiltration basins can be advantageous where secondary use such as playing fields is made of the basin area and where land is available for a sufficiently large open basin. However, an infiltration basin can accumulate low flows, since it has no regular surface outlet. There is no assurance it will remain dry under long-term average conditions unless this objective is taken explicitly into account in basin design. Recent research has provided a basin design equation to accomplish this.

The floor of a basin must remain dry, if a basin is to be considered dry at all. All inflows and outflows during an average year must be considered to occur on the floor area *Af*.

The infiltration rate *I* must not be overestimated, if an adequately dry regime is to be assured. The value used in design calculations must reflect the most limiting of the soil horizons beneath the floor. To take into account the uncertainty surrounding natural soil infiltration rates and reductions due to compaction and sedimentation, a safety factor *Fs* of about 0.5 can be applied in an unvegetated basin to the natural soil *I*.

To protect floor vegetation, the allowable infiltration must be reduced to a rate that will not drown plant roots in saturated soil, according to the following rules: on sloping sites with free lateral drainage, *Fs* would be 0.10 to 0.15 if the most limiting soil horizon is 2 feet deep or less, and 0.20 to 0.25 if the horizon is deeper than 2 feet; on level sites with little free lateral drainage, the above values of *Fs* would be reduced by about half.

Development of design equation for dry basins is described in Bruce K. Ferguson, 1989, Role of the Long-term Water Balance in Design of Multiple-purpose Stormwater Basins, *Proceedings of 1989 Conference of Council of Educators in Landscape Architecture.*

The floor area for a dry basin should take all the above factors into account. The bigger the basin, the greater the monthly losses. With hydraulic gradient assumed equal to 1.0, the floor is sized to dispose all inflow, without accumulation of standing water, in any one individual month:

$$Af = Ro / (Et + IFs + DrX - P)$$

where,

Af	=	floor area with outflows equal to inflow, ac.;
Ro	=	runoff from catchment, ac.ft./mo.;
Et	=	evapotranspiration, ft./mo.;
I	=	infiltration, ft./mo.;
Fs	=	safety factor on infiltration rate, no units;
Dr	=	depth of catchment runoff, ft./mo.;
X	=	proportion of basin overlapping catchment area; and
P	=	precipitation, ft./mo.

A negative floor area can be yielded by the above equation, in months when direct precipitation is very large. This means that net losses from the basin are negative, and it is impossible to dispose all inflow no matter what the basin area. In months when this condition occurs the required floor area should be set infinitely large.

Test all 12 months to see which one requires the largest area. The largest of the monthly basin requirements is the minimum dry floor area for your basin.

Additional design criteria are necessary to assure that an adequately dry regime actually occurs on a floor with a given area.

Shape the basin floor so that it is located above the seasonal high water table. It must be graded level in order to assure that all flows are distributed uniformly to take advantage of the full specified floor area. Over time, an alluvial fan of accumulating sediment could cause low-flow runoff to concentrate upon a small portion of the basin floor, where outflows would not operate at their full rates. Hence it may be prudent to construct a floor larger than the minimum required size, reserving a portion of the floor near the inlet for sediment storage.

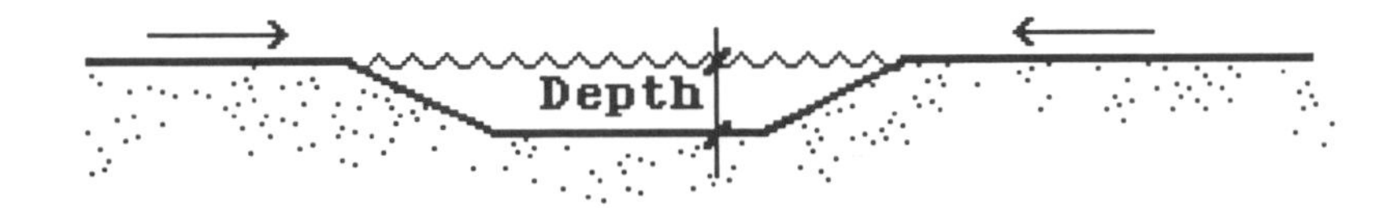

Design-storm ponding time is a concern in multiple-use dry basins. The design storm's runoff must drain out of the basin within a certain number of days, to release the floor for further use. A ponding time from one to seven days might be needed where a basin is actively used such as for scheduled sports. Longer times can be OK where the schedule of use is not so rigid.

Ponding time is controlled by depth in a basin with a broad level floor. The maximum permissible depth to assure an acceptable ponding time can be derived from the basic velocity equation, Velocity = Distance/Time. With depth in feet, infiltration I in ft./day, hydraulic gradient assumed equal to 1.0, and required ponding time in days,

$$\text{Depth} = I \text{ (Ponding time)}$$

The volume of an infiltration basin must be sufficient to hold the runoff volume from a design storm. For a given floor area, the volume is the area times the depth. (Any additional volume resulting from sloping sides could be considered merely extra buffer volume.) Unlike an ephemeral basin, a dry basin is presumed to be dry at the beginning of a design storm, since it is designed to be dry under all ordinary conditions. For a given design-storm volume $Qvol$,

$$Af = Qvol/\text{Depth}$$

Check this area requirement against that derived to assure long-term average dry conditions. Use whichever of the two sizes is larger. A floor that is large enough for the more demanding of the two requirements is sufficient for both.

Near Sarasota, Florida, the Buttonwood Cove residential development infiltrates runoff in a broad, shallow excavated basin planted with St. Augustine grass. Preexisting palm trees were left intact by leaving grass-covered mounds around their roots. A walkway and a picnic table or two are in various corners of the area. Since the area is oversized for the amount of runoff that reaches it there is hardly ever any sign of standing water or wet soil.

Near Orlando, Florida the Maitland Center office development consists of lots densely built up with parking lots and new office buildings. Following a master plan by Post, Buckley, Schuh and Jernigan, each lot has a 30 feet wide perimeter which is ornamented with earth berms, plantings, pedestrian walkways and entry features. Numerous grass basins in the buffer areas infiltrate all the stormwater from the lots. Runoff is delivered from parking lots to the basins by curb cuts, short concrete sluices, and occasional drop inlets with culverts. The emergency overflow from each one drains into the next one downstream. Thus the primary drainage system is down into the soil; the secondary system is across the surface. The secondary system can be traced to its end in a natural lake at the site's lowest elevation.

On Long Island, many of Nassau and Suffolk Counties' "recharge" basins function effectively as dry basins. Although the landscape designs of these basins seldom anticipated multiple use, children from local residences in fact use some of them for recreation. When large enough they make functionally perfect ball fields, since the floor is level and a rolling ball cannot escape over a basin's walls. The location of the basins amidst often dense residential areas, and their presence as the only open space in certain neighborhoods, invited this inadvertent multiple use.

Summary of Process:
Water harvesting
1. Dry infiltration basin

Design-storm ponding:

1. Obtain design-storm volume *Qvol* in ac.ft. from your estimate on page 50.

2. Obtain maximum design-storm ponding time, in days, from local standards.

3. Obtain soil infiltration rate *I* in ft/day from page 75.

4. Compute maximum basin depth in ft.,

$$\text{Depth} = I/\text{Ponding time}.$$

5. Compute floor area *Af* to infiltrate design storm in ac.,

$$Af = Qvol/\text{Depth}$$

Monthly runoff inflow:

6. Obtain monthly runoff *Ro* in ac.ft./mo. from estimate on page 68, 69, 71 or 72.

Monthly net losses:

7. Obtain monthly soil infiltration *I* in ft./mo. from estimate on page 76.

8. Specify infiltration safety factor to protect vegetation *Fs*, no units, from the ranges listed on page 156.

9. Compute monthly allowable infiltration *Ip* in ft./mo.,

$$Ip = I\ Fs$$

10. Obtain monthly runoff depth *Dr* in ft./mo. from estimate on page 68, 69, 71 or 72.

11. Estimate proportion *X* of basin overlapping catchment area (no units), from a site map.

12. Compute monthly loss of runoff depth *Ldr* in ft./mo.,

$$Ldr = Dr\ X$$

13. Obtain monthly turf evapotranspiration *Et* in ft./mo. from water-balance estimate on page 76.

14. Obtain monthly direct precipitation *P* in ft./mo. from page 68, 69, 71 or 72.

15. Sum net losses in ft./mo.,

$$\text{Net losses} = Ip + Ldr + Et - P$$

Long-term dry basin area:

16. Compute monthly floor area *Af* in ac. with outflow equal to inflow,

$$Af = Ro/\text{Net losses}.$$

17. Identify the largest of the floor areas required for the design storm or any of the individual months as the minimum design size.

Water Harvesting Exercise: *1. Dry basin*

Site 1

Design-storm volume Qvol (from p. 50) = ________ac.ft.; Maximum design-storm ponding time = ________days
Infiltration rate *K* (p. 75) = ______ft./day; Design-storm depth = *K*/Ponding time = _____ft.; Min. area *Af* = *Qvol*/Depth = _____ac.
Proportion *X* of basin overlapping catchment area (decimal fraction, from site map) = ________
Infiltration factor *Fs* for protecting vegetation on basin floor (decimal fraction) = ________

	Jan	Feb	Mar	Apr	May	Jun	Jul	Aug	Sep	Oct	Nov	Dec	Annual
Monthly runoff inflow:													
Runoff *Ro*, ac.ft./mo. (from page 68 or 71)	___	___	___	___	___	___	___	___	___	___	___	___	___
Allowable infiltration loss:													
Soil infiltration *I*, ft./mo. (from page 76)	___	___	___	___	___	___	___	___	___	___	___	___	___
Allowable infiltration *Ip*, ft./mo. = *I Fs*	___	___	___	___	___	___	___	___	___	___	___	___	___
Runoff depth loss:													
Runoff depth *Dr*, ft./mo. (from page 68 or 71)	___	___	___	___	___	___	___	___	___	___	___	___	___
Runoff depth loss *Ldr* = *Dr X*	___	___	___	___	___	___	___	___	___	___	___	___	___
Net losses:													
Turf evapotranspiration *Et*, ft./mo. (from page 76)	___	___	___	___	___	___	___	___	___	___	___	___	___
Direct precipitation *P*, ft./mo. (from page 68 or 71)	___	___	___	___	___	___	___	___	___	___	___	___	___
Net losses, ft./mo. = *Ip* + *Ldr* + *Et* - *P*	___	___	___	___	___	___	___	___	___	___	___	___	___
Dry basin area:													
Min. dry floor area *Af*, ac. = *Ro* / net losses	___	___	___	___	___	___	___	___	___	___	___	___	___

Water Harvesting Exercise:
1. Dry basin

Site 2

Design-storm volume Qvol (from p. 50) = ______ ac.ft.; Maximum design-storm ponding time = ______ days
Infiltration rate *K* (p. 75) = ______ ft./day; Design-storm depth = *K*/Ponding time = ______ ft.; Min. area *Af* = *Qvol*/Depth = ______ ac.
Proportion *X* of basin overlapping catchment area (decimal fraction, from site map) = ______
Infiltration factor *Fs* for protecting vegetation on basin floor (decimal fraction) = ______

	Jan	Feb	Mar	Apr	May	Jun	Jul	Aug	Sep	Oct	Nov	Dec	Annual
Monthly runoff inflow:													
Runoff *Ro*, ac.ft./mo. (from page 69 or 72)	___	___	___	___	___	___	___	___	___	___	___	___	___
Allowable infiltration loss:													
Soil infiltration *I*, ft./mo. (from page 76)	___	___	___	___	___	___	___	___	___	___	___	___	___
Allowable infiltration *Ip*, ft./mo. = *I Fs*	___	___	___	___	___	___	___	___	___	___	___	___	___
Runoff depth loss:													
Runoff depth *Dr*, ft./mo. (from page 69 or 72)	___	___	___	___	___	___	___	___	___	___	___	___	___
Runoff depth loss *Ldr* = *Dr X*	___	___	___	___	___	___	___	___	___	___	___	___	___
Net losses:													
Turf evapotranspiration *Et*, ft./mo. (from page 76)	___	___	___	___	___	___	___	___	___	___	___	___	___
Direct precipitation *P*, ft./mo. (from page 69 or 72)	___	___	___	___	___	___	___	___	___	___	___	___	___
Net losses, ft./mo. = *Ip* + *Ldr* + *Et* - *P*	___	___	___	___	___	___	___	___	___	___	___	___	___
Dry basin area:													
Min. dry floor area *Af*, ac. = *Ro* / net losses	___	___	___	___	___	___	___	___	___	___	___	___	___

Dry-basin Exercise
Discussion of Results

1. Estimate the area of each dry basin as a proportion of its catchment area (*Af/Ac*). Which site requires the larger basin? What site and climatic conditions contribute to the requirement of such a large basin?

2. In your judgment, is the required area of either dry basin so large as to be unreasonable in terms purely of land use? What approaches to hydraulic function and site design could you take to overcome this difficulty?

3. What different *types of stormwater effects* are dealt with in each of the two functions, dry-basin water harvesting and stormwater infiltration? What different *types of objectives* are applied to them?

4. Should the additional function of design-storm infiltration necessarily apply to every site where base flows are retained in a dry basin? What other approaches could you take to meeting a combination of requirements for dry-basin water harvesting with control of stormwater's downstream impacts?

5. Which requires the larger floor, a dry average regime or length of ponding time for a design storm. What site conditions, climatic conditions (precipitation and evapotranspiration) and design constraints or assumptions contributed to the requirement of such a large basin? How?

6. Using the floor area derived for a dry basin, route monthly average flows through the basin using the procedures described in the appendix on *Open Ephemeral Basins*. Did the equation on page 157 correctly predict that no water would accumulate in the basin? Is there any excess area in the basin over that necessary to prevent accumulation of standing water? If so, what factor(s) in the equation on page 157 contributed to such "oversizing"?

7. Estimate annual infiltration through your basin by routing it through the procedure for open ephemeral basins in Appendix F. Add up the site's evapotranspiration and infiltration over the year. If the hydrologic objective of the basin is to recharge water into subsurface soil moisture and groundwater, do you think it has accomplished this to a satisfactory degree? How could the basin be revised to increase the proportion that infiltrates?

Wet Basins

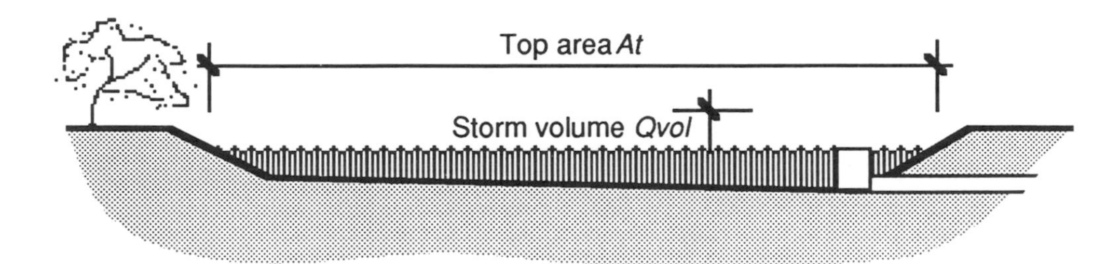

Wet regimes can be advantageous where a permanent pond or wetland is needed for wildlife, aesthetics, recreation, water quality control or recirculated cooling water.

The runoff water supply to a pond maintains the water level and provides throughflow to prevent stagnancy. Runoff has been used frequently to supply landscape ponds, although seldom with quantitative planning to assure adequate supplies. Recent research has provided designers the ability to assure that an adequate water supply is in fact available using a pond design equation.

A pond's supply requirement can be estimated as the sum of monthly losses to evaporation and seepage.

The top area At, the area of water surface exposed to the atmosphere, is where evaporation and direct precipitation occur. Their volumes are proportional to At. However, infiltration losses occur through both the bottom and the sides, which are not parallel to At. Nevertheless, in a basin with side slopes less than about 5:1 (20 percent), the side slope is nearly parallel to the level bottom. In such ponds all outflows, including infiltration, occur in proportion to the total top area At.

The infiltration rate I must not be underestimated, if a wet regime is to be assured. Infiltration is a positive loss of water as long as the floor of the basin is above the local water table. To take into account the uncertainty surrounding natural soil infiltration rates, the value of the safety factor Fs applied to infiltration rate should be 1 or greater than 1.

Development of design equation for wet basins is described in Bruce K. Ferguson, 1989, Role of the Long-term Water Balance in Design of Multiple-purpose Stormwater Basins, *Proceedings of 1989 Conference of Council of Educators in Landscape Architecture.*

The top area for a wet basin should take all the above factors into account. The bigger the area, the greater the monthly losses of water. For basins with side slopes less than about 5:1 the following equation yields the maximum supportable pond area in any one individual month:

$$At = Ro \,/\, (El + IFs + DrX - P)$$

where,

At = top area of water surface with outflows equal to inflow, ac.;
Ro = monthly inflow from catchment, ac.ft./mo.;
El = lake evaporation, ft./mo.;
I = infiltration, ft./mo.;
Fs = safety factor on infiltration rate, no units;
Dr = depth of catchment runoff, ft./mo.;
X = proportion of basin overlapping catchment area; and
P = precipitation, ft./mo.

A negative top area can be yielded by the above equation, in months when direct precipitation P is very large. This means that net losses from the basin are negative, and it is impossible to dispose all inflow no matter what the pond area. In months when this condition occurs the supportable top area should be set infinitely large.

Test all 12 months to see which one supports the smallest area. The smallest of the monthly pond areas is the maximum wet top area for your pond. During the other 11 months a water surplus occurs, and excess runoff is lost to surface overflows through the pond's outlet.

For basins with side slopes steeper than 5:1, the above equation can be used if a pond is sealed using such impervious materials as plastic sheets or a bentonite clay blanket. Sealing makes infiltration rate zero, so that the area of side slopes is irrelevant to the result of the design equation. It can be used also if the basin's sides are impervious vertical retaining walls, making the side area irrelevant to basin hydrology.

The rate of outflow during storms is determined by the outlet's design and by flood storage above the permanent pool's surface. Flood storage can be integrated with a limited supportable pond area At by shaping the flood storage volume to fit. With flood storage volume Vs in ac.ft.,

$$\text{Flood storage depth} = Vs/At$$

An outlet can then be selected which releases the desired outflow rate at the specified water depth. On some sites, storing a required flood volume over a permanent pool may be be impractical because of an extreme required depth.

Near Columbus, Ohio, at the headquarters of the Chemlawn corporation, runoff is supplied from roofs and lawns to a delicately detailed and lavishly planted landscape pond located prominently near the building. During the planning of the site by James H. Bassett, topographic drainage divides were identified. Parking lots and roadways that might produce pollutants were located outside the pond's catchment area. The catchment was contoured to further segregate it from roads and to deliver an adequate supply of runoff to the pond. The surface outlet from the pond flows freely into a lower pond, farther from the building, where large stormwater flows are detained to meet local flood-control standards.

In Charlotte, North Carolina, the University Place multi-use development is focused on an 11-acre lake centered in the bowl-shaped site. At the upper, narrower end of the lake, walkways, fountains, lighting and plazas front on hotels, restaurants and shops of the development's commercial center, making the water surface the unifying theme of an active urban space. At the outer, broader end of the lake, multi-family housing fronts along the lake edge. Since the lake occupies a large part of its watershed, it must be resupplied in some summer months with well water.

In Suffolk County, Long Island, the New York Department of Transportation's Centereach infiltration basin drains a 69-acre highway, commercial and wooded area, with sandy soils and about 10 percent impervious cover. The water table is about 40 feet deep. The basin's half-acre floor is lined with sheets of impervious polyvinyl chloride to form an artificial pond, stocked with fish and aquatic vegetation. Beneath the liner is a gallery of perforated pipes. When the pond level rises during storms, excess water flows through a filter box into the pipes, and infiltrates. The overlying pond improves water quality before it infiltrates, and provides a wetland habitat for wildlife.

In Waterloo, Iowa, the 15-acre office site of Northwest Mortgage supplies runoff to a 1-acre permanent pool prominently located between the wings of the building. The pool, designed as part of the site development by the Durrant Group, provides the building's cooling water by use of an array of low recirculating spray fountains. An edging of gravel and placed boulders accommodates up to two feet of water-level fluctuation for flood storage. The design storm's 62-cfs inflow is reduced to 5 cfs by a drop-inlet weir so as not to overload an existing 12-inch city storm sewer. A well resupplies the pond in some summer months, when rapid natural evaporation is aggravated by high cooling load on the spray fountains. The fountain heads are on risers to allow the cooling system to function even when the storm storage is highest.

Summary of Process
2. Permanent pool

Monthly runoff inflow:

1. Obtain monthly runoff Ro in ac.ft./mo. from estimate on page 68, 69, 71 or 72.

Losses:

2. Obtain monthly lake evaporation El in ft./mo. from water-balance estimate on page 76.

3. Obtain monthly soil infiltration I if ft./mo. from estimate on page 76.

4. Specify infiltration safety factor Fs, no units, equal to or greater than 1.

5. Compute monthly allowable infiltration Ip in ft./mo.,

$$Ip = I\,Fs$$

6. Obtain monthly runoff depth Dr in ft./mo. from estimate on page 68, 69, 71 or 72.

7. Estimate proportion X of basin overlapping catchment area (no units), from a site map.

8. Compute monthly loss of runoff depth Dr in ft./mo.,

$$Ldr = Dr\,X$$

9. Obtain monthly direct precipitation P in ft./mo. from page 68, 69, 71 or 72.

Supportable pool area if sealed:

10. Sum monthly relevant net losses in ft./mo.,

$$\text{Net losses} = El + Ldr - P$$

11. Compute monthly supportable top area At in ac.,

$$At = Ro/\text{Net losses}$$

12. Identify the smallest of the monthly areas as the maximum area for your pond.

Supportable pool area if unsealed:

13. Sum monthly relevant net losses in ft./mo.,

$$\text{Net losses} = El + I + Ldr - P$$

14. Compute monthly supportable top area At in ac.,

$$At = Ro/\text{Net losses}$$

12. Identify the smallest of the monthly areas as the maximum area for your pond.

Water Harvesting Exercise:
Permanent pool, side slopes ≤ 5:1

Site 1

Proportion X of pond overlapping catchment area (decimal fraction, from site map) = ________
Infiltration safety factor Fs = ______ (no units)

	Jan	Feb	Mar	Apr	May	Jun	Jul	Aug	Sep	Oct	Nov	Dec	Annual
Runoff inflow:													
Runoff Ro, ac.ft./mo. (from page 68 or 71)	___	___	___	___	___	___	___	___	___	___	___	___	___
Losses:													
Lake evaporation El, ft./mo. (from page 76)	___	___	___	___	___	___	___	___	___	___	___	___	___
Infiltration I, ft./mo. (from page 76)	___	___	___	___	___	___	___	___	___	___	___	___	___
Allowable infiltration Ip, ft./mo. $= I\ Fs$	___	___	___	___	___	___	___	___	___	___	___	___	___
Runoff depth Dr, ft./mo. = (from page 68 or 71)	___	___	___	___	___	___	___	___	___	___	___	___	___
Runoff depth loss Ldr, ft./mo., $= Dr\ X$	___	___	___	___	___	___	___	___	___	___	___	___	___
Direct precipitation P, ft./mo. (from page 68 or 71)	___	___	___	___	___	___	___	___	___	___	___	___	___
Supportable pool area if sealed:													
Net losses from sealed pond, ft./mo. $= El + Ldr - P$	___	___	___	___	___	___	___	___	___	___	___	___	___
Max. supportable top area At, ac. $= Ro$ / net losses (∞ if neg.)	___	___	___	___	___	___	___	___	___	___	___	___	___
Supportable pool area if unsealed:													
Net losses from unsealed pond, ft./mo. $= El + I + Ldr - P$	___	___	___	___	___	___	___	___	___	___	___	___	___
Max. supportable top area At, ac. $= Ro$ / net losses (∞ if neg.)	___	___	___	___	___	___	___	___	___	___	___	___	___

Water Harvesting Exercise:
Permanent pool, side slopes ≤ 5:1

Site 2

Proportion X of pond overlapping catchment area (decimal fraction, from site map) = ________
Infiltration safety factor Fs = ______ (no units)

	Jan	Feb	Mar	Apr	May	Jun	Jul	Aug	Sep	Oct	Nov	Dec	Annual
Runoff inflow:													
Runoff Ro, ac.ft./mo. (from page 69 or 72)	___	___	___	___	___	___	___	___	___	___	___	___	___
Losses:													
Lake evaporation El, ft./mo. (from page 76)	___	___	___	___	___	___	___	___	___	___	___	___	___
Infiltration I, ft./mo. (from page 76)	___	___	___	___	___	___	___	___	___	___	___	___	___
Allowable infiltration Ip, ft./mo. $= I\ Fs$	___	___	___	___	___	___	___	___	___	___	___	___	___
Runoff depth Dr, ft./mo. = (from page 69 or 72)	___	___	___	___	___	___	___	___	___	___	___	___	___
Runoff depth loss Ldr, ft./mo., $= Dr\ X$	___	___	___	___	___	___	___	___	___	___	___	___	___
Direct precipitation P, ft./mo. (from page 69 or 72)	___	___	___	___	___	___	___	___	___	___	___	___	___
Supportable pool area if sealed:													
Net losses from sealed pond, ft./mo. $= El + Ldr - P$	___	___	___	___	___	___	___	___	___	___	___	___	___
Max. supportable top area At, ac. $= Ro$ / net losses (∞ if neg.)	___	___	___	___	___	___	___	___	___	___	___	___	___
Supportable pool area if unsealed:													
Net losses from unsealed pond, ft./mo. $= El + I + Ldr - P$	___	___	___	___	___	___	___	___	___	___	___	___	___
Max. supportable top area At, ac. $= Ro$ / net losses (∞ if neg.)	___	___	___	___	___	___	___	___	___	___	___	___	___

Permanent Pool Exercise
Discussion of Results

1. Estimate the area of each permanent pool as a proportion of its catchment area (*Af/Ac*). Which site is able to support the larger pool? What site and climatic conditions contribute to the support of such a large pool?

2. In your judgment, is the supportable pool area on either site so small as to be unreasonable in terms of land use and site design? What approaches to hydraulic function and site design could you take to overcome this difficulty?

3. Considering all the possible effects of stormwater, is a bigger volume of runoff following development necessarily good or necessarily bad? What types of site-specific factors could make a large stormwater volume an advantage or a disadvantage?

4. Add up your pond's total annual lake evaporation in ft./mo., and multiply by your pond area to get annual *El* volume. This is a "consumptive" use in the sense that water that is lost to the atmosphere is no longer available for use on the site or in the region. This consumption does not occur on your site until the pond is constructed. Who benefits from this use? Read up on the water-rights laws in your area. Under what conditions would a developer be permitted to install your pond and initiate this consumptive use?

Landscape Irrigation

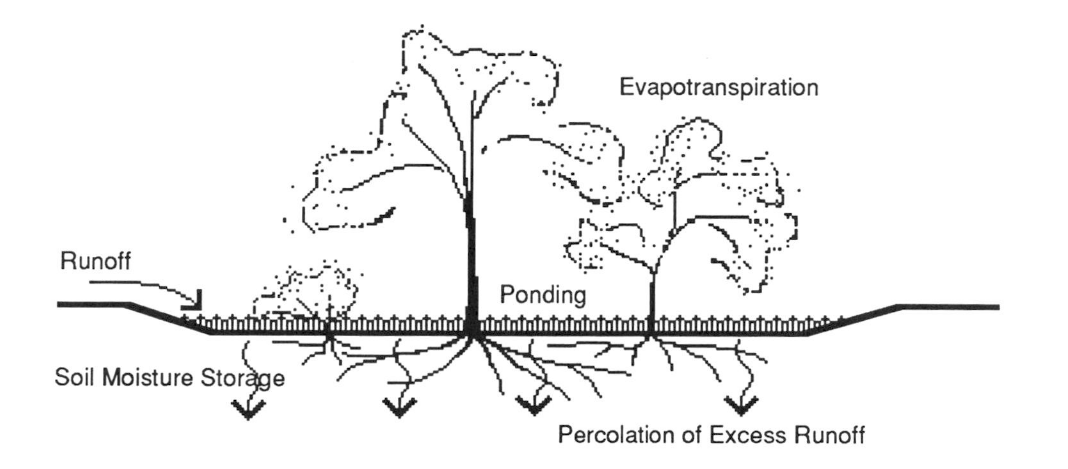

Landscape irrigation is another way to put harvested water to use. Such "runoff irrigation" is being developed notably in Arizona and Colorado, where dry climates cause great demands for irrigation water. Irrigated plantings have included lawns, gardens, shrub areas and fruit and shade trees

Runoff of adequate quality to be used directly by many types of plants is generated by roofs, lawns and paved areas with light traffic.

A planting's irrigation requirement can be estimated by conventional procedures such as the *potential evapotranspiration* method or the *Blaney-Criddle* method, which relate plant evapotranspiration to monthly weather factors such as temperature and solar radiation.

The simplest runoff irrigation technique is to direct runoff into planted areas without formal storage or treatment. Water-loving trees and shrubs are rooted in the bottoms and sides of shallow, level earth basins where runoff ponds up and infiltrates. The initial cost is low. If the particular plant species in the basin cannot tolerate prolonged flooding, apply the dry-basin water balance procedure beginning on page 153 to make sure the soil will not be saturated under long-term average conditions. To see how much irrigation water will need to be artificially supplied during dry months, run your proposed basin through the routing for open ephemeral infiltration basins in Appendix F. Subtract each month's actual evapotranspiration indicated by the routing from the plants' monthly irrigation requirement; any shortfall will need to be supplied by an artificial irrigation system if the plants are to be well maintained.

An irrigation system's *efficiency* is the ratio (between 0 and 1) of amount of water delivered effectively to plants to total amount of water applied, taking into account seepage out of the root zone and other losses. Given a total amount applied, the amount that can be used by plants is only the efficiency times the amount applied. Given a plant water demand, the total amount that needs to be applied is the demand divided by the efficiency. A runoff irrigation system's efficiency might be similar to that of flood irrigation, which is about 0.8.

Efficiency could be improved and irrigable area enlarged with further expense by adding a storage basin and irrigation pumps and pipes.

Storage can balance times of rainstorms with times of water demand. Storage can be in open landscape ponds with edges design to tolerate frequent fluctuation in water level. The necessary size of a storage basin can be evaluated by cumulatively summing its monthly inflows and outflows over a year. If it does not overflow while supplying the use adequately in each month, the largest amount in storage during any month of the year is the minimum required volume.

Pumps and pipes can deliver water to plants when they need it. Each type of irrigation system (flood, sprinkler, drip, etc.) has a more or less predictable efficiency rating.

Whether to install an artificial irrigation system for permanent water supply in addition to an irregular supply of direct runoff is a site-specific question and is related to, among other things, the sensitivity to drought of the type of plants in the basin. With an automatic irrigation system, a moisture sensor such as a rain gauge or tensiometer can be used to delay the normal irrigation schedule during and for some time after each storm, until the soil dries out and watering is in fact needed again. The amount of water saved by that delay may be significant since almost half of some cities' water is used for landscape irrigation.

Collecting and storing water for irrigation affects storm flows the same way infiltration does. If 100 percent of the design-storm runoff is collected then downstream *Qvol* becomes zero and *Qp* is eliminated. Some of the water applied for irrigation may reappear in groundwater or surface streams after passing through slow processes of use and subsurface storage. At that time the runoff will be part of stream and groundwater low flows, not the original storm flows.

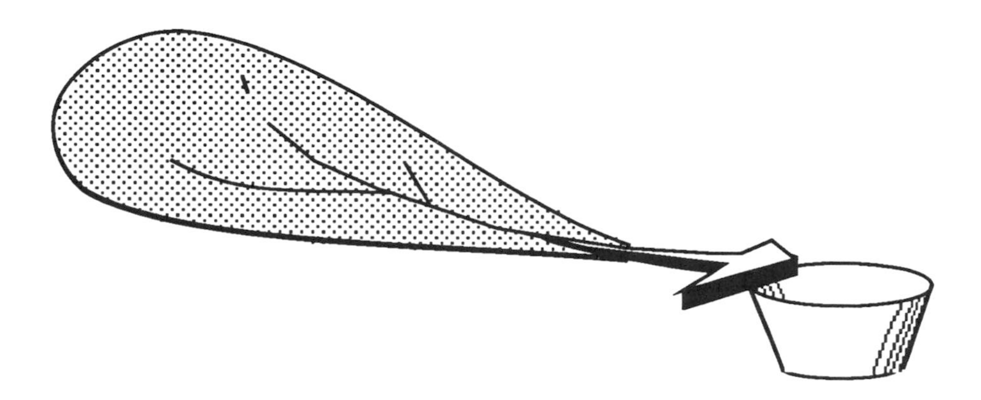

Near Denver, Colorado the Greenwood Plaza office park captures runoff from rainfall and excess landscape irrigation in several small ponds prominently located at road intersections and building facades. Capturing runoff helps satisfy local stormwater control standards. Landscaping with rocks around pond edges is designed to tolerate fluctuations in water level. Pumps regularly move the water into ponds nearest the areas that need irrigation; at the proper time further pumps push the water into irrigation pipes for delivery to plants.

Summary and Commentary

It is ironic that modern cities tend to shunt *away* the rain that falls on them while they import fresh water from distant streams and reservoirs.

Conservation of the water we are already using may be the cheapest, fastest and most direct way to extend the supplies of water-short cities and regions.

By making positive use of on-site rainfall and runoff, water harvesting promises to reduce need for imported water.

Unlike other methods of controlling stormwater, water harvesting turns the liability of stormwater impact into the asset of water supply. It offers the incentive of positive supply to constructing and properly maintaining a stormwater control facility.

The quality of harvested water depends on the land uses in the catchment area. Dumpster pads, service areas where trash will be handled, and heavily used parts of roads and parking lots can be some of the most intense generators of pollutants. Roofs, turf that is not over-fertilized, and paved areas with only light traffic generate relatively clean runoff which could be suitable for pond or irrigation water supplies. The site can be laid out and the catchment graded to direct runoff from pollutant generators away from the harvest area. Runoff of questionable quality could be run through a settling basin or vegetated filter area before delivery to a sensitive application.

Using runoff during storms for positive supply requires an integration of contouring, planting, drainage and general site layout that seldom occurs where stormwater is not reclaimed.

Nevertheless, hydraulics need not dictate urban design. The results of water-harvesting calculations are basin areas and depths. They are parameters for site design, not the design itself. As long as you meet these few quantitative criteria, you are free to design the site any way that is needed in terms of planting, contouring, materials, etc.

Epilogue

Water can be made to go through specific processes as it passes over and through a development site: conveyance, storm detention, extended detention, infiltration and water harvesting.

Quantitative methods are available for evaluating stormwater features and assuring that a proposed design will accomplish its objective. When there is more than one way to estimate something, such as "rational" and SCS storm runoff estimates, and you are uncertain which one is better, apply each of them, and use the result that is safer and more conservative. Where a facility has to meet a combination of objectives, such as storm detention and extended detention, solve for each of them separately, then use the design that is most limiting. When the variability of nature makes you uncertain of the values to use in an equation, such as infiltration rates or cover factors, apply a safety factor to make your assumptions conservative and assure that your design will work.

Qualitative differences among hydrologic functions are more fundamental than quantitative calculations for governing future flows and storages. Before it is worth your time to argue about a quantitative, relative distinction, you must first explicitly choose the *kind* of process for which you are designing. That choice will stay with you through all the quantitative calculations for carrying it out. The choice of qualitative approach determines the *types* of environments water will move toward, the *types* of processes it will pass through, and the *types* of effects it will have.

The choice of hydrologic function is a site-specific decision. You have to decide how you want your landscape to work: the kind of environment you want to create.

We are at a fantastic point in history. Research on new methods of controlling, conserving and using stormwater is continuing. Our ability to apply even the relatively old methods in site-specific design develops as we gain more experience. New methods, new understandings and new capabilities are *changing the way we build cities.*

Society wants help with water. It is a resource on which everyone can agree. Through appropriate design for controlling and using runoff we can help to protect, conserve and supply water in all of its qualitative and quantitative aspects, both in the urban infrastructure and in the natural environment.

Designers are in a position to have a real impact on the world. Designers *form* land development. They control *the surface of the earth.*

The surface of the earth is the thin membrane through which all rainfall must pass. Through it all water is partitioned into the specific surface, subsurface and artificial flows of which the hydrology of a development site or an urban region is composed. By knowledgeably laying out, grading, draining and planting that surface and building structures upon it, we can direct the locations and rates of water's flows and storages through all parts of social and natural systems.

Bibliography

Abt, Steven R., and others, 1988, Resistance to Flow over Riprap in Steep Channels, *Water Resources Bulletin* vol. 24, no. 6, pages 1193-1200.

Adams, L.W., and others, 1984, Public Attitudes toward Urban Wetlands for Stormwater Control and Wildlife Enhancement, *Wildlife Society Bulletin* vol. 12, no. 3, pages 299-303.

American Iron and Steel Institute, 1980, *Modern Sewer Design*, Washington: American Iron and Steel Institute.

American Public Works Association Research Foundation, 1981, *Urban Stormwater Management*, Special Report No. 49, Chicago: American Public Works Association.

American Society of Civil Engineers, 1987, *Ground Water Management, Third Edition*, New York: American Society of Civil Engineers.

Aron, Gert, David J. Wall, Elizabeth L. White and Christopher N. Dunn, 1987, Regional Rainfall Intensity - Duration - Frequency Curves for Pennsylvania, *Water Resources Bulletin*, vol. 23, no. 3, pages 479-485.

Aronson, D.A. and G.E. Seaburn, 1974, *Appraisal of Operating Efficiency of Recharge Basins on Long Island, New York,* Water-Supply Paper 2001-D, Washington: U.S. Geological Survey.

Asano, Takashi (editor), 1985, *Artificial Recharge of Groundwater,* Stoneham, Massachusetts: Butterworth Publishers.

Bianchi, W.C. and Dean C. Muckel, 1970,

Ground-Water Recharge Hydrology, ARS 41-161, Washington: U.S. Agricultural Research Service.
Blaney, Harry F. and Wayne D. Criddle, 1962, *Determining Consumptive Use and Irrigation Water Requirements,* Technical Bulletin No. 1275, Washington: U.S. Agricultural Research Service.
Brune, Gunnar M., 1953, Trap Efficiency of Reservoirs, *Transactions of the American Geophysical Union* vol. 34, no. 3, pages 407-418.
Cavacas, Alan, 1985, Regional Watershed Management — Some Unanswered Questions, pages 109-120 of *Proceedings, 1985 International Symposium on Urban Hydrology, Hydraulic Infrastructures and Water Quality Control*, University of Kentucky, Lexington, Kentucky, July 23-25, 1985.
Cedergren, H.R.,1977, *Seepage, Drainage, and Flow Nets, Second Edition*, New York: Wiley.
Cedergren, H.R., K.J. O'Brien and J.A. Arman, 1972, *Guidelines for the Design of Subsurface Drainage Systems for Highway Structural Systems*, Report No. FHWA-RD-72-30, Washington: U.S. Federal Highway Administration, Office of Research and Development.
Chow, Ven Te, and others, 1988, *Applied Hydrology*, New York: McGraw-Hill.
Chow, Ven Te, 1959, *Open-channel Hydraulics*, New York: McGraw-Hill.
College of Agriculture, (no date), *Water Harvesting Systems,* Tucson, Arizona: University of Arizona College of Agriculture.
Debano, Leonard F. and Burchard H. Heede, 1987, Enhancement of Riparian Ecosystems with Channel Structures, *Water Resources Bulletin* vol. 23, no. 3, pages 463-470.
Debo, Thomas N., 1985, Personal Computers and Stormwater Management Programs, in *Computer Applications in Water Resources,* edited by Harry C. Torno, New York: American Society of Civil Engineers.
Debo, Thomas N., 1980, Stormwater Management: The Concept and Its Application, *Water Resources Bulletin* vol. 16, no. 4, pages 654-660.
Debo, Thomas N., 1976, The Complete Drainage Program — More Than an Ordinance, *Water Resources Bulletin* vol. 12, no. 1, pages 109-121.
Debo, Thomas N. and Bruno O. Ulrich, 1977, Stormwater Program Is Model for Others, *Public Works* vol. 108, no. 7.
Debo, Thomas N. and Holly Ruby, 1982, Detention Basins, An Urban Experience, *Public Works* January, 1982, pages 42-43, 93.
Debo, Thomas N. and George E. Small, 1989, Detention Storage — Its Design and Use, *Public Works*, January, 1989, pages 71-72.
Debo, Thomas N. and George E. Small, 1989, Hydrologic Calibration: the Forgotten Aspect of Drainage Design, Public Works Feburary, 1989, pages 58-59.
Debo, Thomas N., Michael Elliott and Arthur C. Nelson, 1988, County Begins to Develop an Impact Fee Program, *Public Works* November, 1988, pages 44-46.
Dendy, F.E., 1974, Sediment Trap Efficiency of Small Reservoirs, *Transactions of the American Society of Agricultural Engineers* vol. 17, pages 898-901 and 908.
Dingman, S. Lawrence, 1984, *Fluvial Hydrology,* New York: Freeman.
Dokoolian, N.K., Jr., V.E. Petrucci, J.E. Ayars, C.D. Clary and R.A. Schoneman, 1987, Artificial Ground Water Recharge by Flooding during Grapevine Dormancy, *Water Resources Bulletin* vol 23, no. 2, pages 307-311.
Driver, N.E., and G.D. Tasker, 1989, *Techniques for Estimation of Storm-runoff Loads, Volumes, and Selected Constituent Concentrations in Urban Watersheds in the United States*, Water-Resources Investigations Report 88-191, Denver: U.S. Geological Survey, Water Resources Division.
Dunne, Thomas and Luna B. Leopold, 1978, *Water in Environmental Planning,* San Francisco: Freeman.
Eagleman, Joe R., 1976, *The Visualization of Climate,* Lexington, Massachusetts: Lexington Books (D.C. Heath and Company).
Eagleson, Peter S., 1970, *Dynamic Hydrology*, New York: McGraw-Hill.

Ebbert, J.C., and R.J. Wagner, 1987, Contributions of Rainfall to Constituent Loads in Storm Runoff from Urban Catchments, *Water Resources Bulletin* vol. 23, no. 5, pages 867-871.

Engstrand, Daniel, 1983, *Retention Ponds: Analysis and Design of Ponds without Outlets*, St. Paul, Minnesota: Minnesota Dept. of Transportation.

Farr, E., and W. C. Henderson, 1986, *Land Drainage*, London: Longman.

Ferguson, Bruce K., 1990 (in press), Role of Long-Term Water Balance in Design of Multiple-Purpose Stormwater Basins, *Proceedings*, 1989 Conference of Council of Educators in Landscape Architecture.

Ferguson, Bruce K., 1990 (in press), Role of Long-Term Water Balance in Management of Stormwater Infiltration, *Journal of Environmental Management.*

Ferguson, Bruce K., 1988, Using Water Effectively, chapter 3 of *Irrigation, Handbook of Landscape Architectural Construction vol. 3*, Washington: Landscape Architecture Foundation.

Ferguson, Bruce K., 1987, Urban Stormwater Harvesting: Applications and Hydraulic Design, *Journal of Environmental Management* vol. 25, pages 71-79.

Ferguson, Bruce K., 1987, Water Conservation Methods in Urban Landscape Irrigation: An Exploratory Overview, *Water Resources Bulletin* vol. 23, no. 1.

Ferguson, Bruce K., 1987, Environmental Patterns of Water Management, *Journal of Environmental Systems* vol. 16, no. 3, pages 161-178.

Ferguson, Bruce K., 1985, Land Environments of Water Resource Management, *Journal of Environmental Systems* vol. 13, no. 3, pages 291-312.

Ferguson, Bruce K., 1983, Landscape Hydrology: A Unified Guide to Water-Related Design, Pages 11-21 of *Proceedings, The Landscape: Critical Issues and Resources,* 1983 Conference of Council of Educators in Landscape Architecture, Utah State University, Logan, Utah.

Ferguson, Bruce K., 1983, Infiltration Basins, *Landscape Architecture* vol. 73, no. 6, pages 89-91.

Ferguson, Bruce K., 1981, Erosion and Sedimentation Control in Regional and Site Planning, *Journal of Soil and Water Conservation* vol. 35, no. 4, pages 199-204.

Ferguson, Bruce K., 1981, *Controlling Stormwater Impact,* LATIS Series, Washington: American Society of Landscape Architects.

Ferguson, Bruce K., 1978, Erosion and Sedimentation Control in Site Master Planning, *Journal of Soil and Water Conservation* vol. 33, no. 4, pages 167-172.

Ferguson, Bruce K., and Philip W. Suckling, 1989, Variations in Precipitation and Runoff in an Urbanizing Watershed, pages 145-148 of of *Proceedings of the 1989 Georgia Water Resources Conference*, Kathryn J. Hatcher, ed., Athens: University of Georgia, Institute of Natural Resources.

Fichtel, C., and L.W. Adams, 1982, *Permanent Ponds for Urban Stormwater Control and Wildlife Enhancement in Allegheny County, Pennsylvania*, Columbia, Maryland: Urban Wildlife Research Center.

Field, Richard, 1985, Urban Runoff: Pollution Sources, Control, and Treatment, *Water Resources Bulletin* vol.. 21, no. 2, pages 197-206.

Fink, Dwayne H. and William J. Ehrler, 1984, The Runoff Farming Agronomic System: Applications and Design Concepts, *Hydrology and Water Resources in Arizona and the Southwest* vol. 14, pages 33-40.

Fisher, G.T., and B.G. Katz, 1988, *Urban Stormwater Runoff, Selected Background Information and Techniques for Problem Solving, with a Baltimore, Maryland, Case Study*, Water-Supply Paper 2347, Washington: U.S. Geological Survey.

Florida Non-Point Source Management Division, (no date), *Example Design of a Multi-Purpose Detention / Filtration Facility*, Tallahassee, Florida: Florida Department of Environmental Resources.

Florida Non-Point Source Management Division, (no date), *Standards and Specifications for Stormwater Retention Basins*, Tallahassee, Florida: Florida Department of

Environmental Resources.

Florida Non-Point Source Management Division, (no date), *Standards and Specifications for Underdrains and Stormwater Filtration Systems*, Tallahassee, Florida: Florida Department of Environmental Resources.

Frasier, Gary W., 1980, Harvesting Water for Agricultural, Wildlife, and Domestic Uses, *Journal of Soil and Water Conservation,* vol. 35, no. 3, pages 125-128.

Frasier, Gary W. and Lloyd E. Myers, 1983, *Handbook of Water Harvesting,* Agriculture Handbook 600, Washington: U.S. Department of Agriculture.

Frederick, Ralph H., and others, 1977, *Five to 60 Minute Precipitation and Frequency for the Eastern and Central United States*, Technical Memorandum NWS HYDRO-35, Silver Spring, Md.: National Weather Service, Office of Hydrology.

Fresno Metropolitan Flood Control District, 1977, *Infiltration Drainage Design for Highway Facilities*, Fresno, Calif.: Fresno Metropolitan Flood Control District.

Friedman, J., 1985, *Wetlands Hydrology and Sedimentation, Implications for the Design and Management of Wetland Preserves*, Seattle, Washington: The Nature Conservancy.

Georgia Crushed Stone Association, 1987, *Stormwater Management with Infiltration Basins*, Atlanta: Georgia Crushed Stone Association.

Godfrey, K., and others, 1985, *Ecological Considerations in Wetlands Treatment of Municipal Wastewaters*, New York: Van Nostrand Reinhold.

Halverson, Howard G., David R. DeWalle and William E. Sharpe, 1984, Contribution of Precipitation to Quality of Urban Storm Runoff, *Water Resources Bulletin* vol. 20, no. 6, pages 859-864.

Hannon, Joseph B., 1980, *Underground Disposal of Storm Water Runoff, Design Guidelines Manual,* FHWA-TS-80-218, Washington: U.S. Federal Highway Administration.

Harrington, B.W., 1986, *Feasibility and Design of Wet Ponds to Achieve Water Quality Control*, Annapolis: Maryland Department of Natural Resources.

Hartigan, John P., 1983, Watershed-Wide Approach Significantly Reduces Local Stormwater Management Costs, *Public Works* December 1983, pages 34-37.

Heaney, James P., and Wayne C. Huber, 1984, Nationwide Assessment of Urban Runoff Impact on Receiving Water Quality, *Water Resources Bulletin* vol. 20, no. 1, pages 35-42.

Heath, Ralph C., 1984, *Ground-Water Regions of the United States*, Water-Supply Paper 2242, Washington: U.S. Geological Survey.

Heath, Ralph C., 1983, *Basic Ground-Water Hydrology*, Water-Supply Paper 2220, Washington: U.S. Geological Survey.

Hershfield, David M., 1961, *Rainfall Frequency Atlas of the United States,* Technical Paper 40, Washington: U.S. Department of Commerce, Weather Bureau.

Hewlett John D., 1982, *Principles of Forest Hydrology*, Athens: University of Georgia Press.

Huisman, L., and T.N. Olsthoorn, 1983, *Artifical Groundwater Recharge*, Boston: Pitman Advanced Publishing Program.

Jenkins, D. and F. Pearson, 1978, *Feasibility of Rainwater Collection Systems in California,* Contribution No. 173, Berkeley: University of California, California Water Resources Center.

Jensen, Marvin E., editor, 1973, *Consumptive Use of Water and Irrigation Water Requirements,* New York: American Society of Civil Engineers.

Jones, O.R., D.W. Goss and A.D. Schneider, 1981, Management of Recharge Basins on the Southern High Plains, *Transactions of the American Society of Agricultural Engineers* vol. 24, no. 4, pages 977-980 and 987.

Kibler, D.F., editor, 1982, *Urban Stormwater Hydrology*, Water Resources Monograph No. 7, Washington: American Geophysical Union.

King, Horace Williams, and Ernest F. Brater, 1954, *Handbook of Hydraulics for the Solu-*

tion of Hydraulic Problems, Fourth edition, New York: McGraw-Hill.

King County Resource Planning Section, 1986, *Viability of Freshwater Wetlands for Urban Surface Water Management and Nonpoint Pollution Control: An Annotated Bibliography*, Seattle: State of Washington Department of Ecology.

Kohler, M.A., T.J. Nordenson and W.E. Fox, 1955, *Evaporation from Pans and Lakes*, Technical Paper 38, Washington: U.S. Weather Bureau.

Kozlowski, James C., 1985, Retention Pond Drowning Case Studies, *Parks & Recreation,* March 1985, pages 38-44 and 95.

Ku, Henry F.H. and Dale L. Simmons, 1986, *Effect of Urban Stormwater Runoff on Ground Water Beneath Recharge Basins on Long Island, New York*, Water-Resources Investigations Report 85-4088, Syosset, New York: U.S. Geological Survey.

Kuo, Chin Y., Kelly A. Cave and G.V. Loganathan, 1988, Planning of Urban Best Management Practices, *Water Resources Bulletin* vol. 24, no. 1, pages 125-132.

LaBaugh, James W., 1986, Wetland Ecosystem Studies from a Hydrologic Perspective, *Water Resources Bulletin* vol. 22, no. 1, pages 1-10.

Lamoreaux, W.W., 1962, Modern Evaporation Formulae Adapted to Computer Use, *Monthly Weather Review* vol. 90, pages 26-28.

Leopold, Luna B., 1974, *Water, A Primer,* San Francisco: Freeman.

Leopold, Luna B., 1968, *Urban Hydrology for Land Planning*, Circular 554, Washington: U.S. Geological Survey.

Lindsey, Greg, 1988, *A Survey of Stormwater Utilities*, Annapolis: Maryland Department of the Environment, Stormwater Management Administration.

Livingston, Eric, and others, 1988, *The Florida Development Manual: A Guide to Sound Land and Water Management,* Tallahassee: Florida Department of Environmental Regulation.

Mallard, G.E., 1980, *Microorganisms in Stormwater — a Summary of Recent Investigations*, Open-File Report 80-1198, Washington: U.S. Geological Survey.

Martin, Edward H., 1988, Effectiveness of an Urban Runoff Detention Pond-Wetlands System, *Journal of Environmental Engineering* vol. 114, no. 4, pages 810-827.

Maryland Water Resources Administration, 1986, *Minimum Water Quality Objectives and Planning Guidelines for Infiltration Practices*, Annapolis: Maryland Department of Natural Resources.

Maryland Water Resources Administration, 1987, *Guidelines for Constructing Wetland Stormwater Basins*, Annapolis: Maryland Department of Natural Resources.

Maryland Water Resources Administration, 1985, *Inspector's Guidelines Manual for Stormwater Management Infiltration Practices,* Annapolis: Maryland Department of Natural Resources.

Maryland Water Resources Administration, 1984, *Standards and Specifications for Infiltration Practices,* Annapolis: Maryland Department of Natural Resources.

McCuen, Richard H., 1982, *A Guide to Hydrologic Analysis Using SCS Methods,* Englewood Cliffs, New Jersey: Prentice-Hall.

McCuen, Richard H., Multicriterion Stormwater Management Methods, *Journal of Water Resources Planning and Management* vol. 114, no. 4, pages 414-431.

Meyer, Adolph F., 1961, Storage Requirements for Beneficial Use, Chapter 9 of *National Engineering Handbook, Section 4, Hydrology*, Washington: U.S. Soil Conservation Service.

Miller, David H., 1977, *Water at the Surface of the Earth, an Introduction to Ecosystem Hydrodynamics*, New York: Academic.

Miller, John F., 1963, *Probable Maximum Precipitation and Rainfall-Frequency Data for Alaska*, Technical Paper No. 47, Washington: U.S. Weather Bureau.

Miller, John F, and others, 1973 *Precipitation-Frequency Atlas of the United States*, Atlas 2, Washington: National Oceanic and Atmospheric Admistration.

Miller, T.L., and S.W. McKenzie, 1978, *Analysis of Urban Storm-water Quality from Sev-*

en Basins near Portland, Oregon, Open-File Report 78-662, Washington: U.S. Geological Survey.

Miller, W.P., and M.K. Baharuddin, 1986, Relationship of Soil Dispersability to Infiltration and Erosion of Southeastern Soils, *Soil Science* vol. 142, no. 4, pages 235-240.

Morgan, Arthur Ernest, 1951, *The Miami Conservancy District,* New York: McGraw-Hill.

National Stone Association, 1982, *Quarried Stone for Erosion and Sediment Control,* Washington: National Stone Association.

Nightingale, Harry I., 1987, Water Quality Beneath Urban Runoff Water Management Basins, *Water Resources Bulletin* vol. 23, no. 2, pages 197-205.

Nightingale, Harry I., 1987, Accumulation of As, Ni, Cu, and Pb in Retention and Recharge Basins Soils from Urban Runoff, *Water Resources Bulletin* vol. 23, no. 4, pages 663-672.

Nightingale, Harry I., and W.C. Bianchi, 1977, *Environmental Aspects of Water Spreading for Ground-water Recharge*, Technical Bulletin no. 1568, Washington: U.S. Department of Agriculture, Science and Education Administration.

Nightingale, Harry I., and W.C. Bianchi, 1973, Groundwater Recharge for Urban Use, *Ground Water* vol. 11, no. 6, page 36-43.

Nix, Stephan J., and Ting-Kuei Tsay, 1988, Alternative Strategies for Stormwater Detention, *Water Resources Bulletin* vol. 24, no. 3, pages 609-614.

Normann, Jerome M., Robert J. Houghtalen and William J. Johnston, 1985, *Hydraulic Design of Highway Culverts,* Hydraulic Design Series No. 5, Washington: U.S. Federal Highway Administration.

Novitzki, R.P., 1982, *Hydrology of Wisconsin Wetlands,* Information Circular 40, Washington: U.S. Geological Survey.

O'Hare, Margaret, Deborah M. Fairchild, Paris A. Hajali and Larry W. Canter, 1986, *Artificial Recharge of Ground Water, Status and Potential in the Contiguous United States,* Chelsea, Michigan: Lewis Publishers.

Pensyl, L. Kenneth, and Paul F. Clement, 1987, *Results of the State of Maryland Infiltration Practices Survey,* Annapolis: Maryland Department of Natural Resources, Water Resources Administration, Stormwater Management Division.

Peterson, Frank L., and David R. Hargis, Subsurface Disposal of Storm Runoff, 1973, *Journal Water Pollution Control Federation* vo. 45, no. 8, pages 1663-1670.

Pettyjohn, Wayne A., (no date), *Introduction to Artificial Groundwater Recharge,* Ada, Oklahoma: U.S. Environmental Protection Agency, Robert S. Kerr Environmental Research Laboratory.

Pham, C.H., H.G. Halverson and G.M. Heisler, 1978, *Precipitation and Runoff Water Quality from an Urban Parking Lot and Implications for Tree Growth,* Note NE-253, Broomall, Pennsylvania: U.S. Forest Service Research.

Pluhowski, E.J., and A.G. Spinello, 1978, Impact of Sewarage Systems on Stream Base Flow and Ground-water Recharge on Long Island, New York, U.S. Geological Survey *Journal of Research* vol. 6, no. 2, pages 263-271.

Poertner, H.G., 1974, *Practices in Detention of Urban Stormwater Runoff,* Special Report No. 43, Chicago: American Public Works Association.

Rawls, W.J., D.L. Brakensiek, and K.E. Saxton, 1982, Estimation of Soil Water Properties, *Transactions of the American Society of Agricultural Engineers* vol. 25, no. 5, pages 1316-1320 and 1328.

Ruffner, James A. and Frank E. Bair, 1984, *The Weather Almanac,* Detroit: Gale Research Company.

Sartor, J.D., G.B. Boyd and F.J. Agardy, 1974, Water Pollution Aspects of Street Surface Contaminants, *Journal of the Water Pollution Control Federation* vol. 46, no. 3, pages 458-467.

Schueler, Thomas R., 1987, *Controlling Urban Runoff: A Practical Manual for Planning and Designing Urban Best Management Practices,* Metropolitan Washington Council of Governments.

Seaburn, G.E., 1969. *Effects of Urban Develop-*

ment on Direct Runoff to East Meadow Brook, Nassau County, Long Island, New York, Professional Paper 627-B, Washington: U.S. Geological Survey.

Seaburn, G.E. and D.A. Aronson, 1974, *Influence of Recharge Basins on the Hydrology of Nassau and Suffolk Counties, Long Island, N.Y.*, Water-Supply Paper 2031, Washington: U.S. Geological Survey.

Simmons, Dale L. and Richard J. Reynolds, 1982. Effects of Urbanization on Base Flow of Selected South-Shore Streams, Long Island, New York. *Water Resources Bulletin* vol. 18, no. 9, pages 797-805.

Skopp, J., and T.C. Daniel, 1978, A Review of Sediment Predictive Techniques as Viewed from the Perspective of Nonpoint Pollution Management, *Environmental Management* vol. 2, no. 1, pages 39-53.

Smith, T.W., R.R. Peter, R.E. Smith and E.C. Shirley, 1969, *Infiltration Drainage of Highway Surface Water*, Research Report 6328201, Sacramento, California: California Department of Transportation.

Snyder, W.M., and A.W. Thomas, 1987, Patterns of Watershed Monthly Runoff, *Water Resources Bulletin* vol. 23, no. 6, pages 1133-1140.

Solley, Wayne B., Edith B. Chase and William B. Mann IV, 1983, *Estimated Use of Water in the United States in 1980*, Circular 1001, U.S. Geological Survey.

Stockdale, Erik C., 1986, *The Use of Wetlands for Stormwater Managment and Nonpoint Pollution Control: A Review of the Literature*, Seattle: State of Washington Department of Ecology.

Striegl, Robert G., 1987, Suspended Sediment and Metals Removal from Urban Runoff by a Small Lake, *Water Resources Bulletin* vol. 23, no. 6, pages 985-996.

Sykes, Robert D., 1988, Surface Water Drainage, pages 133-221 of *Handbook of Landscape Architectural Construction, Volume 1,* editor Maurice Nelischer, Washington: Landscape Architecture Foundation.

Sykes, Robert D., 1988, Channels and Ponds, pages 239-317 of *Handbook of Landscape Architectural Construction, Volume Two,* editor Maurice Nelischer, Washington: Landscape Architecture Foundation.

Tasker, Gary D., and Nancy E. Driver, 1988, Nationwide Regression Models for Predicting Urban Runoff Water Quality at Unmonitored Sites, *Water Resources Bulletin* vol. 24, no. 5, pages 1091-1101.

Toro Company, 1966, *Rainfall-Evapotranspiration Data, United States and Canada*, Minneapolis, Minnesota: Toro Company.

U.S. Bureau of Reclamation, 1974, *Design of Small Dams, 2nd Edition,* Washington: U.S. Bureau of Reclamation.

U.S. Federal Highway Administration, 1975, *Design of Stable Channels with Flexible Linings,* Hydraulic Engineering Circular No. 15, Washington: U.S. Federal Highway Adminsitration.

U.S. Federal Highway Administration, 1975, *Hydraulic Design of Energy Dissipators for Culverts and Channels,* Hydraulic Engineering Circular No. 14, Washington: U.S. Federal Highway Adminsitration.

U.S. Federal Highway Administration, 1973, *Design Charts for Open-Channel Flow,* Hydraulic Design Series No. 3, Washington: U.S. Federal Highway Adminsitration.

U.S. Federal Highway Administration, 1973, *Design of Roadside Drainage Channels,* Hydraulic Design Series No. 4, Washington: U.S. Federal Highway Adminsitration.

U.S. Soil Conservation Service, 1972, *National Engineering Handbook, Section 4, Hydrology,* SCS/ENG/NEH-4, Washington: U.S. Soil Conservation Service.

U.S. Soil Conservation Service, 1986, *Urban Hydrology for Small Watersheds*, Technical Release 55, 2nd Edition, Washington: U.S. Soil Conservation Service.

U.S. Soil Conservation Service, 1982, *Ponds — Planning, Design, Construction,* Agriculture Handbook 590, Washington: U.S. Soil Conservation Service.

U.S. Soil Conservation Service, 1977, *Design of Open Channels,* Tech. Release 25, Washington: U.S. Soil Conservation Service.

U.S. Soil Conservation Service, 1956, *Hydraulics,* National Engineering Handbook Section 5, Washington: U.S. Soil Conservation

Service.
U.S. Weather Bureau, 1962, *Rainfall-Frequency Atlas of the Hawaiian Islands*, Technical Paper No. 43, Washington: U.S. Weather Bureau.
U.S. Weather Bureau, 1961, *Generalized Estimates of Probable Maximum Precipitation and Rainfall-Frequency Data for Puerto Rico and Virgin Islands*, Technical Paper No. 42, Washington: U.S. Weather Bureau.
U.S. Weather Bureau, 1955, *Rainfall Intensity-Duration-Frequency Curves for Selected Stations in the United States, Alaska, Hawaiian Islands, and Puerto Rico,* Technical Paper No. 25, Washington: U.S. Weather Bureau.
Viessman, Warren, Jr., 1989, Technology, Society, and Water Management, *Journal of Water Resources Planning and Management*, vol. 115, no. 1, pages 48-51.
Walesh, Stuart G., 1989, *Urban Surface Water Management*, New York: Wiley.
Waller, D.H., 1989, Rain Water — an Alternative Source in Developing and Developed Countries, *Water International* vol. 14, pages 27-36.
Walling, D.E., and K.J. Gregory, 1970, The Measurement of the Effects of Building Construction on Drainage Basin Dynamics, *Journal of Hydrology* vol. 2, pages 129-144.
Wanielista, Martin P., 1986, Best Management Practices Overview, pages 314-323 of *Urban Runoff Quality — Impact and Quality Enhancement Technology, Proceedings of an Engineering Foundation Conference*, New England College, Henniker, New Hampshire, June, 1986.
Wanielista, Martin P., Yousef A. Yousef, Bernard L. Golding and Claude L. Cassagnol, 1984, *Stormwater Management Manual*, Tallahassee: Florida Department of Environmental Regulation.
Wanielista, Martin P., 1979, *Stormwater Management, Quantity and Quality,* Ann Arbor: Ann Arbor Science Publishers.
Warner, James W., and others, 1989, Mathematical Analysis of Artificial Recharge from Basins, *Water Resources Bulletin* vol. 25, no. 2, pages 401-411.
Whipple, William, Jr., 1981, Dual Purpose Detention Basins in Storm Water Management, *Water Resources Bulletin* vol. 17, no. 4, pages 642-646.
Whipple, William, Jr., editor, 1975, *Urbanization and Water Quality Control*, Minneapolis: American Water Resources Association.
Whipple, William, Jr., and Joseph V. Hunter, 1981, Settleability of Urban Runoff Pollution, *Journal Water Pollution Control Federation* vol. 53, no. 12, pages 1726-1731.
Whipple, William, Jr., and Joseph V. Hunter, 1979, Petroleum Hydrocarbons in Urban Runoff, *Water Resources Bulletin* vol. 15, no. 4, pages 1096-1105.
Whipple, William, Jr., and others, 1983, *Stormwater Management in Urbanizing Areas*, Englewood Cliffs, NJ: Prentice-Hall.
Whipple, William, Jr., and others, 1978, Runoff Pollution from Multiple Family Housing, *Water Resources Bulletin* vol. 14, no. 2, pages 288-301.
Wigginton, Parker J., Clifford W. Randall and Thomas J. Grizzard, 1986, Accumulation of Selected Trace Metals in Soils of Urban Runoff Swale Drains, *Water Resources Bulletin* vol. 22, no. 1, pages 73-79.
Wigginton, Parker J., Clifford W. Randall and Thomas J. Grizzard, 1983, Accumulation of Selected Trace Metals in Soils of Urban Runoff Detention Basins, *Water Resources Bulletin* vol. 19, no. 5, pages 709-718.
Winter, Thomas C., 1981, Uncertainties in Estimating the Water Balance of Lakes, *Water Resources Bulletin* vol. 17, no. 1, pages 82-115.
Whyte, William H., 1989, Prospect, *Landscape Architecture* vol. 79, no. 6, page 120.
Yousef, Yousef A., and others, 1986, Nutrient Transformations in Retention/Detention Ponds Receiving Highway Runoff, *Journal Water Pollution Control Federation* vol. 58, no. 8, pages 838-844.
Yousef, Yousef A., and others, Fate of Heavy Metals in Stormwater Runoff from Highway Bridges, *Science of the Total Environment* vol. 33, pages 233-244.

Thanks to:

Ted Walker, who encouraged the making of this book and the freedom to do it well, and who has provided a unique service to designers through the continuing succcess of his publishing work;

Bob Nicholls and Darrel Morrison, successive Deans of the the University of Georgia's School of Environmental Design, who have provided an administrative environment for productive activities of the faculty;

A.B. Ferguson, Jr., who helped provide the computer on which this book was created; and

Students of our university courses and professionals who have attended our seminars and short courses, none of whom has had a shortage of useful and penetrating questions to stimulate research and articulation of new ideas.

This book was produced on a Macintosh Plus computer with MacBottom 20Mb external hard drive. Software for writing and page assembly was ReadySet-Go 4.5, after transferring MacWrite files from the book's first edition. For illustrative graphics SuperPaint 2.0, MacDraw II, Cricket Draw and MacAtlas were used with the Helvetica font, after transferring MacPaint and FullPaint files from the first edition. For numerical charts CricketGraph, supported by Excel, was used with the Helvetica font. Camera-ready copy was produced from ReadySetGo files by laser printers at 300 dpi. The finished book occupies 2.7Mb of data, with 2.5 Mb of additional supporting files such as spreadsheet calculations and original graphics.

Appendix A

The exercises in this book are for you to practice applying mathematical models and design procedures.

Contrast between two sites that are different in key respects such as slope, rainfall or hydrologic soil group is, in our experience, very useful for stimulating discussion and understanding of the site conditions that create hydrologic effects and constrain stormwater control facilities.

You can use sites in your region that you are familiar with or that you have come across in your work.

Or you can select from the following pairs of sites. These are hypothetical sites that we have used in our short courses. Although the specific drainage areas are hypothetical, their characteristics such as soil, slope and vegetation are representative of their regions, and actual regional climatic data are used. Some pairs are regionally focused; some leap between different regions of the country. Some contrast local conditions of soil and slope; others contrast different climates.

1. Southeastern Pair

Charleston is in the Coastal Plain physiographic region, which is characterized by level topography and sandy soils overlying major aquifers. Atlanta is in the Piedmont region, which is characterized by low hills and red clayey soil overlying impermeable crystalline rock. Nevertheless the two cities have essentially similar rainfall. When similar land uses are placed in these two regions, contrasts in runoff and required drainage facilities are primarily due to differences in soil and slope. Differences in drainage facilities that result from those conditions are sometimes *not* what one would expect.

	Atlanta	***Charleston***
Location:	Atlanta, Ga.	Charleston, S.C.
Physiography:	Piedmont	Coastal Plain
Design storm recurrence interval:	10 years	10 years
Drainage area:	10 acres	10 acres
Soil type:	Davidson clay loam	Pottsburg loamy sand
Hydraulic length:	1,000 ft.	1,000 ft.
Slope along hydraulic length:	4 %	1 %
10% elevation:	705 ft.	36 ft.
85% elevation:	735 ft.	44 ft.
Existing land use:	Thin forest (pine-oak)	Thin forest (pine-oak)
Proposed land use:	Multi-family residential, about 50% impervious	Multi-family residential, about 50% impervious
Slope of conveyance pipe:	2 %	2 %
Slope of conveyance swale:	2 %	2 %
Swale material:	Grass mixture	Grass mixture
Culvert outlets to:	Grass mixture swale	Grass mixture swale
Maximum depth of water in swale:	2 ft.	2 ft.
Maximum storm detention basin depth:	3 ft.	3 ft.
Minimum residence time in extended detention basin	10 days	10 days
Maximum ponding time in multiple-use dry basin:	7 days	7 days

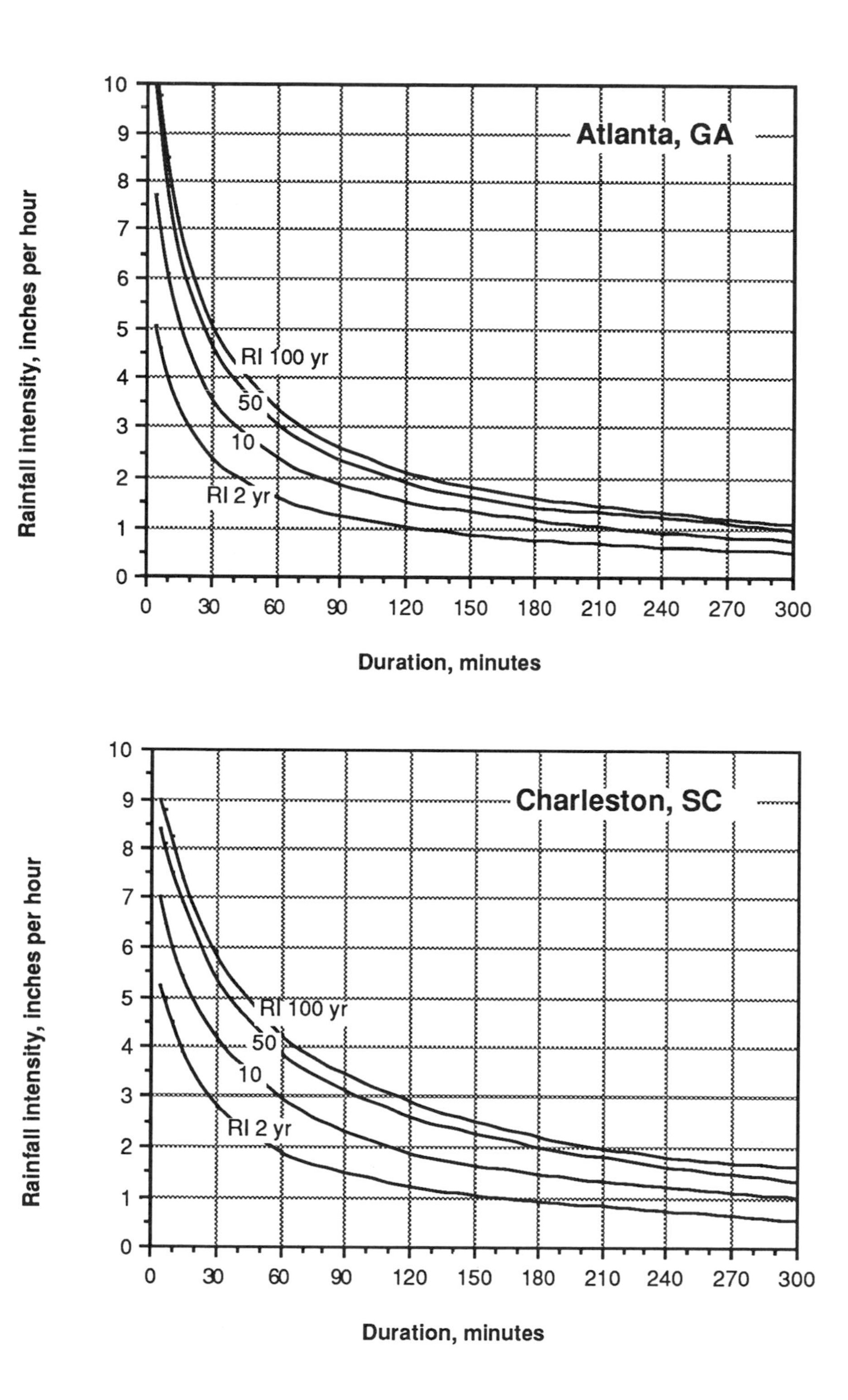

Charts on this page were derived from U.S. Weather Bureau, 1955, *Rainfall Intensity-Duration-Frequency Curves*, Technical Paper No. 25.

2. Northeastern Pair

Atlantic City is on the Coastal Plain, where level topography and sandy soils overlie major aquifers. Chester, near Philadelphia, is in the Piedmont region, where low hills and fine-textured soil overlie impermeable crystalline rock. Nevertheless the two cities have essentially similar rainfall. When similar land uses are placed in these two regions, contrasts in runoff and required drainage facilities are primarily due to differences in soil and slope. Differences in drainage facilities that result from those conditions are sometimes *not* what one would expect.

	Philadelphia	***Atlantic City***
Location:	Chester, Pa.	Atlantic City, N.J.
Physiography:	Piedmont	Coastal Plain
Design storm recurrence interval:	10 years	10 years
Drainage area:	10 acres	10 acres
Soil type:	Glenville silt loam	Evesboro sand
Hydraulic length:	1,000 ft.	1,000 ft.
Slope along hydraulic length:	8 %	1 %
10% elevation:	374 ft.	36 ft.
85% elevation:	434 ft.	44 ft.
Existing land use:	Thin forest (cedar-oak)	Thin forest (pine-oak)
Proposed land use:	Multi-family residential, about 50% impervious	Multi-family residential, about 50% impervious
Slope of conveyance pipe:	2 %	2 %
Slope of conveyance swale:	2 %	2 %
Maximum depth of water in swale:	2 ft.	2 ft.
Swale material:	Grass mixture	Grass mixture
Culvert outlets to:	Grass mixture swale	Grass mixture swale
Maximum storm detention basin depth:	3 ft.	3 ft.
Minimum residence time in extended detention basin	10 days	10 days
Maximum ponding time in multiple-use dry basin:	7 days	7 days

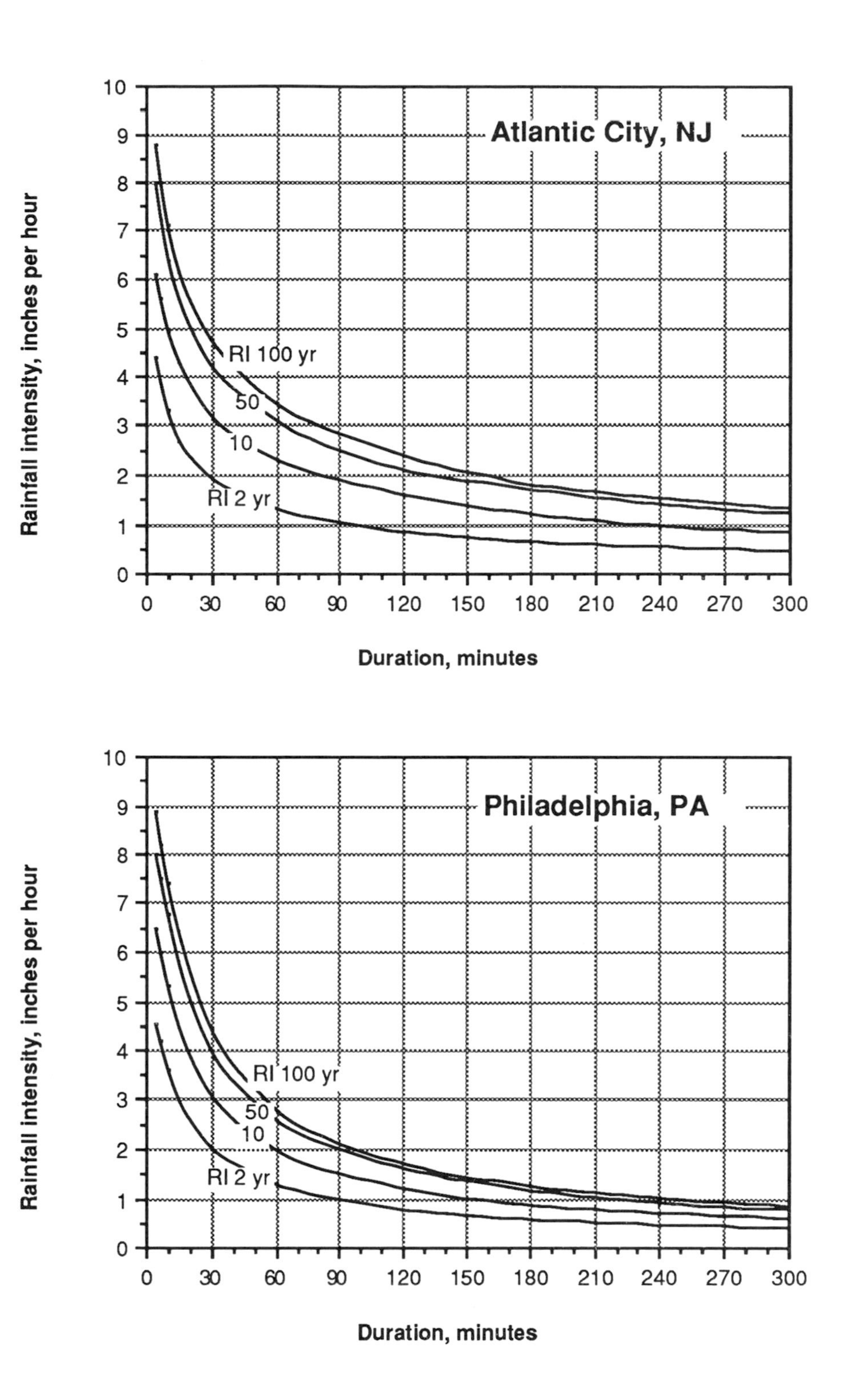

Charts on this page were derived from U.S. Weather Bureau, 1955, *Rainfall Intensity-Duration-Frequency Curves*, Technical Paper No. 25.

2. New York Pair

Glaciation dominates the surface of upstate New York. Ithaca is on the Appalachian Plateau, where till thinly mantles slopes of impervious sedimentary rock. Rochester is on the lake plain to the north, where outwash and groundwater fill the lower elevations. Although the soil textures in the two regions do not differ drastically, if we point out the relatively fine-textured soil on the Plateau, and the relatively coarse-textured soil on the plain, the two sites represent quite a hydrologic contrast. Nevertheless the two locations have essentially similar rainfall, and differences in runoff and required drainage facilities are mostly a result of differences in soil and slope.

	Upland till	*Outwash plain*
Location:	Ithaca, N.Y.	Rochester, N.Y.
Physiography:	Plateau	Lake plain
Design storm recurrence interval:	10 years	10 years
Drainage area:	10 acres	10 acres
Soil type:	Langford channery silt loam	Palmyra gravelly loam
Hydraulic length:	1,000 ft.	1,000 ft.
Slope along hydraulic length:	8 %	2 %
10% elevation:	705 ft.	36 ft.
85% elevation:	735 ft.	44 ft.
Existing land use:	Thin forest (cherry-locust)	Thin forest (cherry-locust)
Proposed land use:	Multi-family residential, about 50% impervious	Multi-family residential, about 50% impervious
Slope of conveyance pipe:	2 %	2 %
Slope of conveyance swale:	2 %	2 %
Length of culvert:	50 ft.	50 ft.
Allowable depth at culvert mouth:	2 ft.	2 ft.
Swale material:	Grass mixture	Grass mixture
Culvert outlets to:	Grass mixture swale	Grass mixture swale
Maximum depth of water in swale:	2 ft.	2 ft.
Maximum storm detention basin depth:	4 ft.	4 ft.
Minimum residence time in extended detention basin	10 days	10 days
Maximum ponding time in multiple-use dry basin:	7 days	7 days

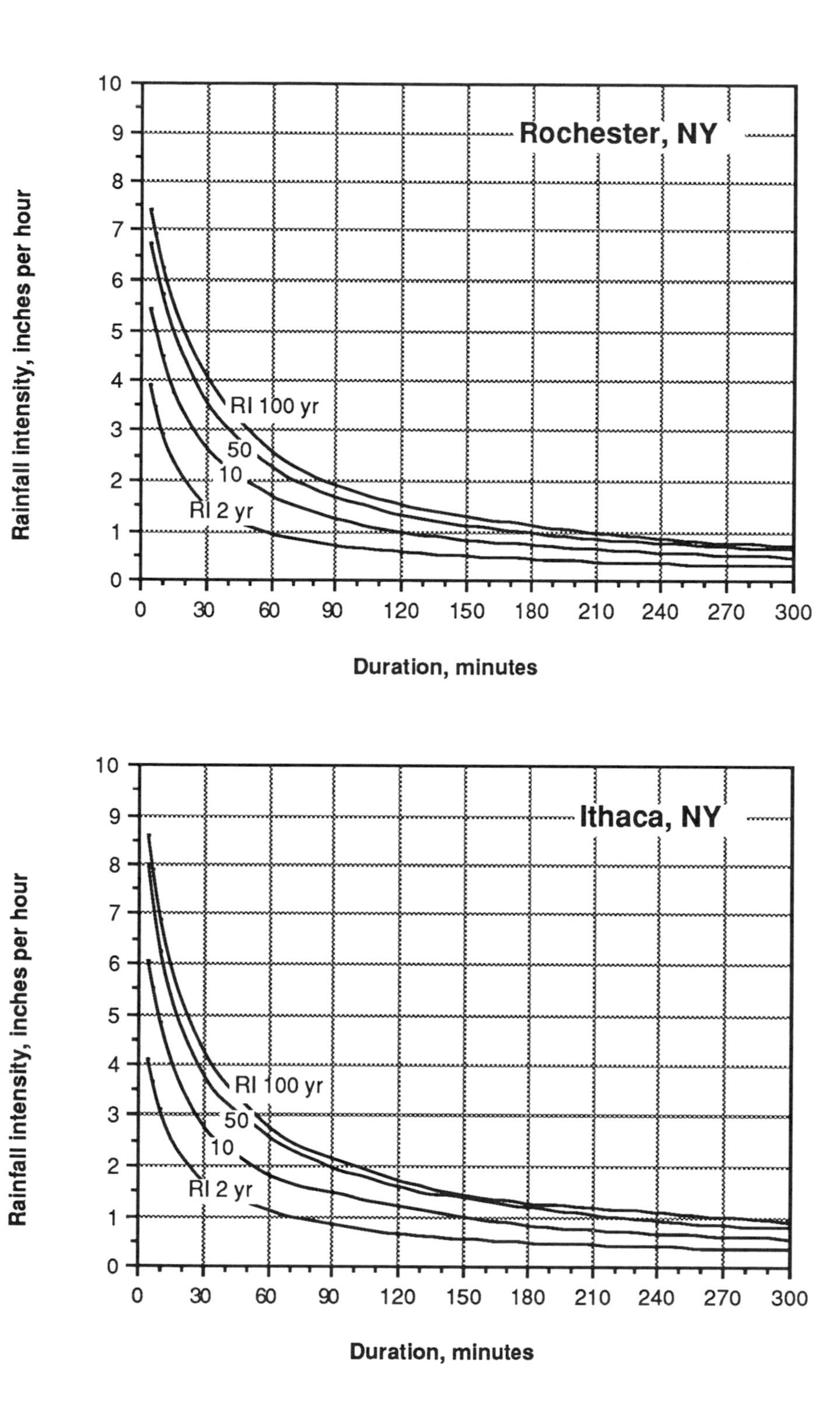

Charts on this page were derived from U.S. Weather Bureau, 1955, *Rainfall Intensity-Duration-Frequency Curves*, Technical Paper No. 25.

2. New Mexico Pair

The powerful relief of the Albuquerque area brings about sharp physiographic divisions. Sloping alluvial fans drape the feet of the mountains, while river terraces and yahoos of the Rio Grande form low-lying level planes. The low rainfall of the Southwest leads to small required drainage facilities, which are convenient from a land use viewpoint. But the aridity limits the potential uses that can be made of harvested water.

	Alluvial fan	*River terrace*
Location:	Albuquerque	Albuquerque
Physiography:	Alluvial fan	River terrace
Design storm recurrence interval:	10 years	10 years
Drainage area:	10 acres	10 acres
Soil type:	Embudo gravelly fine sandy loam	Vinton loamy sand
Hydraulic length:	1,000 ft.	1,000 ft.
Slope along hydraulic length:	5 %	1 %
10% elevation:	705 ft.	36 ft.
85% elevation:	735 ft.	44 ft.
Existing land use:	pinyon-juniper, fair condtion	pinyon-juniper, fair condition
Proposed land use:	Multi-family residential, about 50% impervious	Multi-family residential, about 50% impervious
Slope of conveyance pipe:	2 %	2 %
Slope of conveyance swale:	2 %	2 %
Length of culvert:	50 ft.	50 ft.
Allowable depth at culvert mouth:	2 ft.	2 ft.
Swale material:	Grass mixture	Grass mixture
Culvert outlets to:	Grass mixture swale	Grass mixture swale
Maximum depth of water in swale:	2 ft.	2 ft.
Maximum storm detention basin depth:	4 ft.	4 ft.
Minimum residence time in extended detention basin	10 days	10 days
Maximum ponding time in multiple-use dry basin:	7 days	7 days

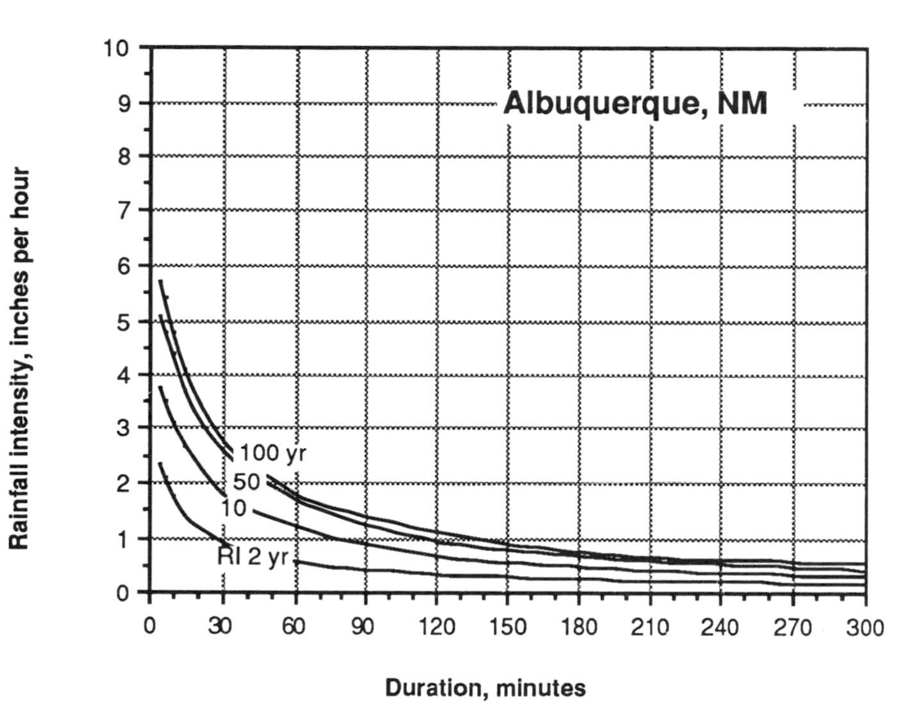

Chart on this page was derived from U.S. Weather Bureau, 1955, *Rainfall Intensity-Duration-Frequency Curves*, Technical Paper No. 25.

3. National Pair

Although Phoenix and Albany represent a contrast of arid and humid climates at opposite corners of the country their *storm* rainfalls are surprisingly similar. In Albany rainstorms come from regularly occurring frontal systems. In Phoenix the big storms tend to be summer thunderstorms stimulated by overheated desert soil. Because of the similarity of storm rainfall their modest contrast in storm runoff is due mostly to differences in soil, topography and vegetation. However major climatic differences show up when the *long-term* water balance is applied.

	Albany	***Phoenix***
Location:	Albany, N.Y.	Phoenix, Ariz.
Physiography:	Glacial lake deposits	Alluvial terrace
Design storm recurrence interval:	10-year storm	10-year storm
Drainage area:	10 acres	10 acres
Soil type:	Schoharie silt loam	Pinal gravelly loam
Hydraulic length:	1,000 ft.	1,000 ft.
Slope along hydraulic length:	3 %	3 %
10% elevation:	251	1,605
85% elevation:	274	1,628
Existing land use:	Old-field woods (pine-hardwood)	Desert shrub (cactus, grasses)
Proposed land use:	Multi-family residential, about 50% impervious	Multi-family residential, about 50% impervious
Slope of conveyance pipe:	2 %	2 %
Slope of conveyance swale:	2 %	2 %
Swale material:	Grass mixture	Grass mixture
Culvert outlets to:	Grass mixture swale	Grass mixture swale
Maximum depth of water in swale:	2 ft.	2 ft.
Maximum storm detention basin depth:	3 ft.	3 ft.
Minimum residence time in extended detention basin	10 days	10 days
Maximum ponding time in multiple-use dry basin:	7 days	7 days

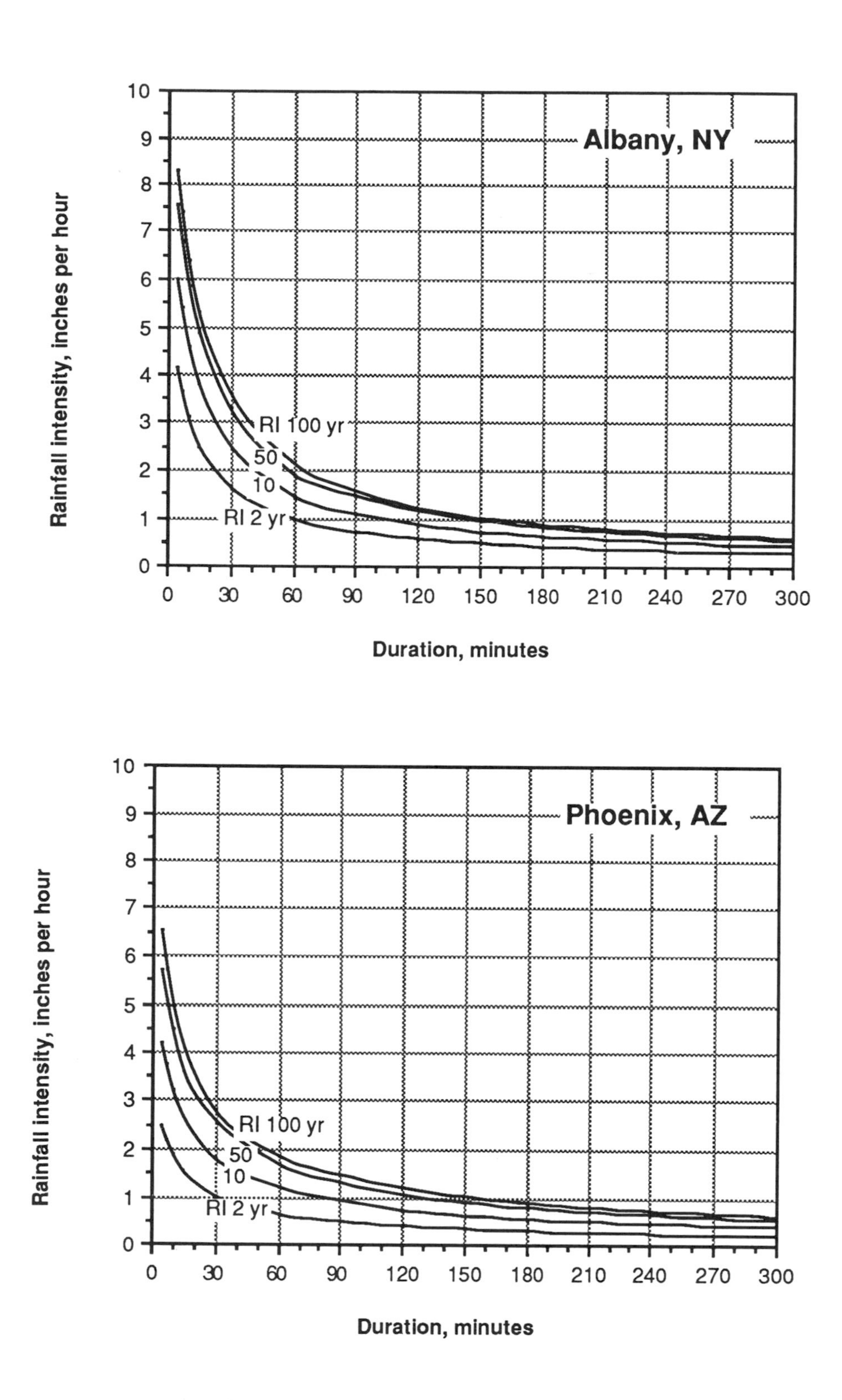

Charts on this page were derived from U.S. Weather Bureau, 1955, *Rainfall Intensity-Duration-Frequency Curves*, Technical Paper No. 25.

3. Texas Pair

Houston and El Paso represent a strong contrast of arid and humid climates at opposite ends of Texas. Houston is on the Coastal Plain near the Gulf of Mexico, receiving a constant flow of moisture from the southern winds. El Paso is in the western desert at the Rio Grande. Nevertheless local sites that are similar in terms of slope and soil hydrologic response can be found in both regions. When similar sites are compared, any differences in stormwater hydrology and management are a result of differences in rainfall and evaporation.

	Houston	***El Paso***
Location:	Houston, Texas	El Paso, Texas
Physiography:	Coastal plain	Alluvial basin
Design storm recurrence interval:	10-year storm	10-year storm
Drainage area:	10 acres	10 acres
Soil type:	Boy loamy fine sand	Hueco loamy fine sand
Hydraulic length:	1,000 ft.	1,000 ft.
Slope along hydraulic length:	1 %	1 %
10% elevation:	88	3,244
85% elevation:	95	3,251
Existing land use:	Oak-pine woods, fair condition	Desert shrub, fair condition
Proposed land use:	Multi-family residential, about 50% impervious	Multi-family residential, about 50% impervious
Slope of conveyance pipe:	2 %	2 %
Slope of conveyance swale:	2 %	2 %
Swale material:	Grass mixture	Grass mixture
Culvert outlets to:	Grass mixture swale	Grass mixture swale
Maximum depth of water in swale:	2 ft.	2 ft.
Maximum storm detention basin depth:	3 ft.	3 ft.
Minimum residence time in extended detention basin	10 days	10 days
Maximum ponding time in multiple-use dry basin:	7 days	7 days

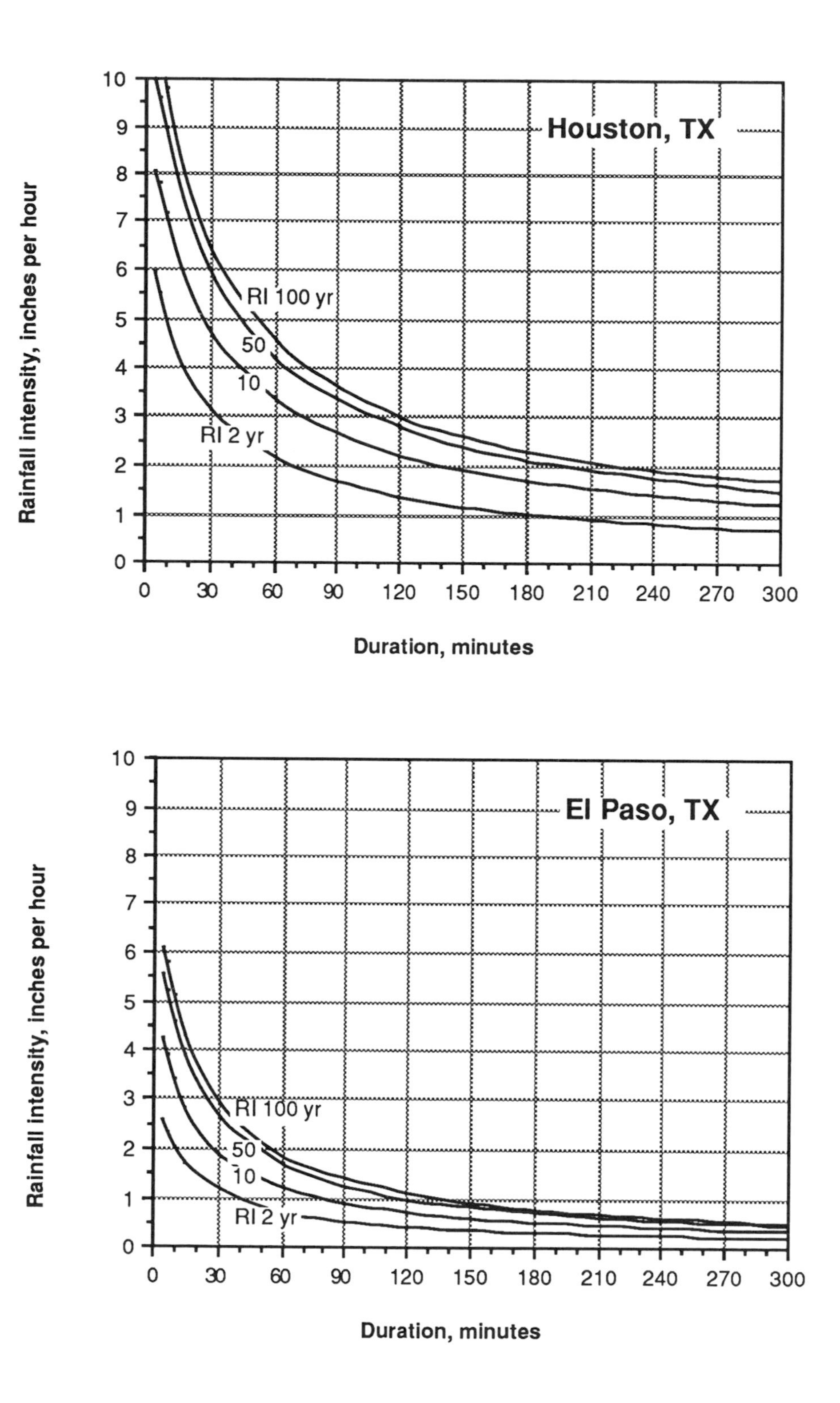

Charts on this page were derived from U.S. Weather Bureau, 1955, *Rainfall Intensity-Duration-Frequency Curves*, Technical Paper No. 25.

In classes we usually divide the participants into two groups, each group doing the exercises for one of the sites in a pair. Thus we generate two sets of solutions which we can compare and discuss.

Appendix B

To estimate runoff using rainfall-runoff models such as the "rational" or the SCS you need to characterize a drainage area's size, land use, soils and other attributes required by specific models.

The first step is to define the drainage area precisely by drawing its boundaries on a site map.

Do not confuse those boundaries with prominent ridgelines or other visible landforms. Drainage area boundaries need to be identified using a very specific analytic process. They may or may not coincide with visible landforms.

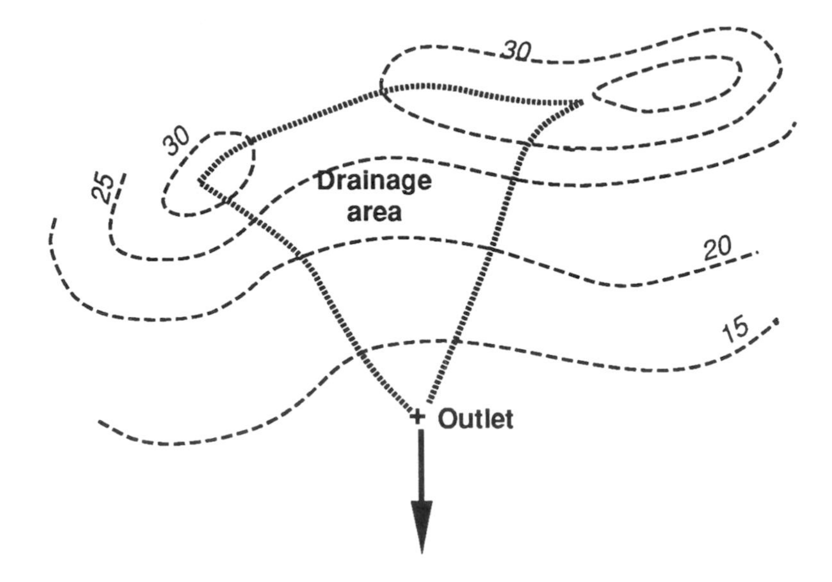

A drainage area is defined by the point in the topography where you need to estimate runoff. Identify the point where you are going to do something important for stormwater control. Only then can you analyze the drainage area of that point. We will call that point the drainage area's outlet. The outlet might be the mouth of a culvert that you have to design, a drainage outlet from a construction area where you are locating a detention basin, or a pond you want to supply with harvested runoff. If it is a pipe, use the pipe's upper end since that is where all the water enters; for a swale or basin use the lower end since that is where the drainage area is largest.

Topographic contour lines tell you where a drainage area's boundary is. Before development follow existing contours. After development follow proposed contours.

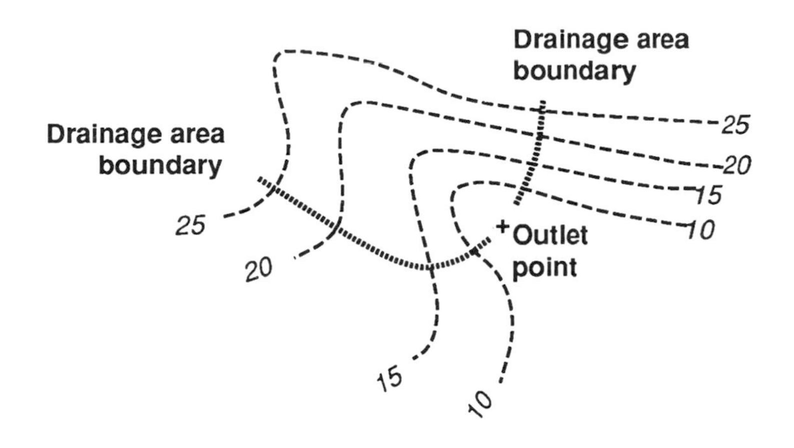

The contours near the outlet indicate two opposite boundaries of a drainage area, converging upon the outlet from uphill. Between the boundaries is the area that is inside the drainage area and drains toward the outlet. Outside the boundaries runoff drains elsewhere.

Draw the boundary on one side starting exactly from the outlet. Move *uphill* from contour to contour by the shortest possible path. Cross each contour at a right angle. On one side of the line you're drawing a running drop of water would flow eventually through the outlet — the area on that side of the line is *inside* the drainage area. On the other side of your line a running drop of water would miss the outlet — that side of your line is *outside* the drainage area.

When you get to a topographic high point such as the top of a hill or knob, stop working on that line for a minute.

Start again at the outlet and go *uphill* in the other available direction, just like before. Again keep going until you reach a high point.

If you end up connecting at the high point with your first line, you are done; your boundary line wraps completely around the watershed.

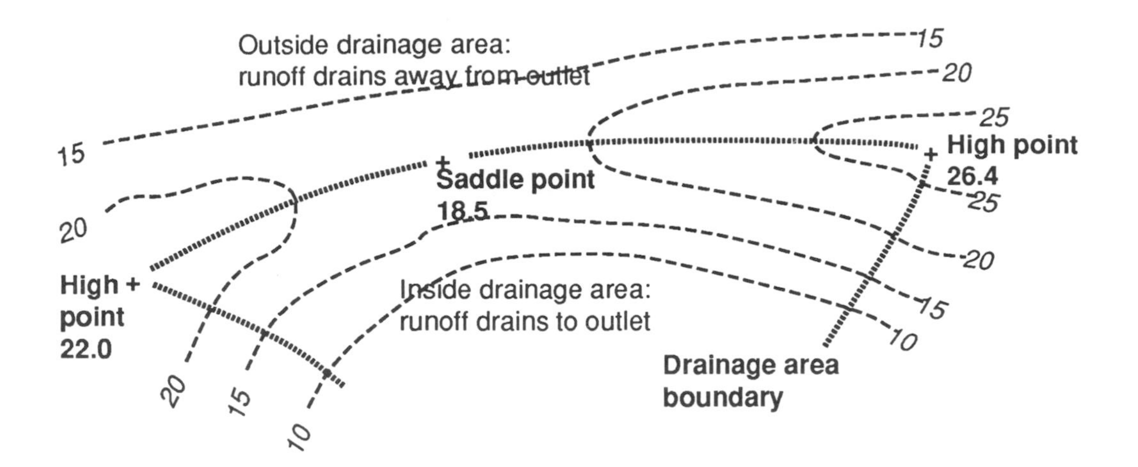

If your lines do not connect yet then your lines have reached two separate topographic knobs. You need to connect the knobs through a saddle point.

A saddle point is a local "low" point between knobs. On two sides of a saddle point the topography goes *up* toward two different high points. On two other sides the topography goes *down* the hillsides. So a saddle point is as low as you can get along the ridgeline connecting two knobs. It is called a saddle point because you can imagine sitting in it with your legs hanging down the two low sides, and the front and rear of the saddle rising up in front and in back of you.

Look for a saddle point near one of your knobs where water would flow toward the drainage area's outlet on one low side, and away from it on the other. Such a saddle is a local low point along the drainage area's boundary. So you can start from it to work up the contours toward each knob, just as you originally did from the outlet.

If necessary continue a succession of lines, from each saddle point to each high point, until they all connect and wrap completely around the area that drains to the outlet.

Structures like buildings and roads are just special cases of the above principles. You are likely to encounter plenty of them when analyzing a drainage area after development. Continue drawing a drainage area boundary from contour to contour as follows.

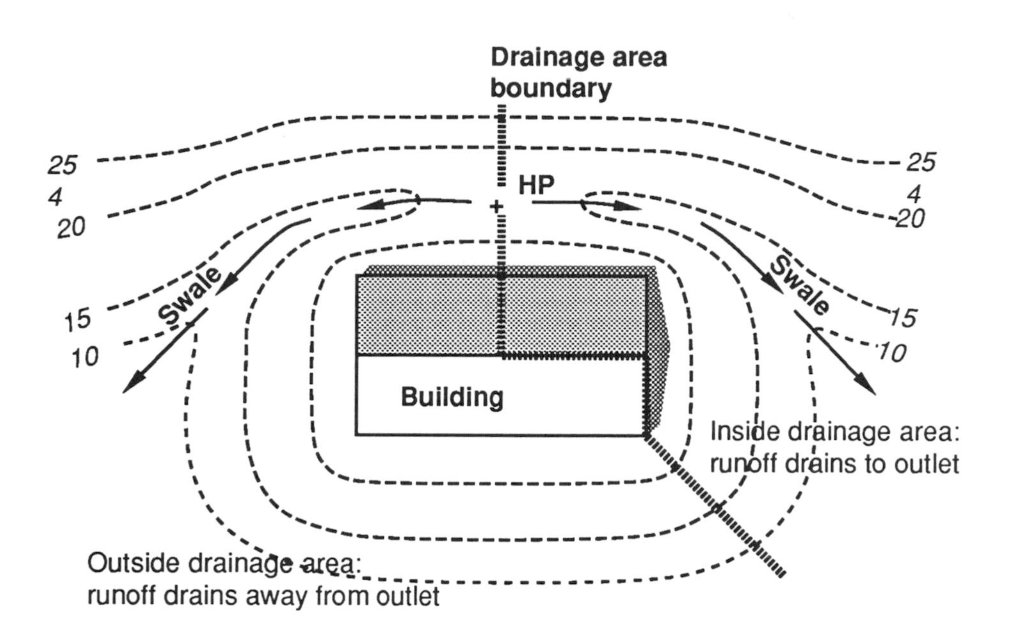

Approach a building from the contours below. Follow the *edge* of the building up to the ridgeline of its roof. The ridgeline is a high *line* in the building's "topography." Now start looking for how to approach the roofline from the other low side, just as you would the top of a knob. The low point is likely to be the high point of a swale on the uphill side of the building. A swale's high point is in fact a topographic saddle point. Start there and work up toward the building's ridgeline, meeting the line you drew before. Then you can continue the drainage area's boundary by working uphill from the swale's high point, as you would from any saddle point.

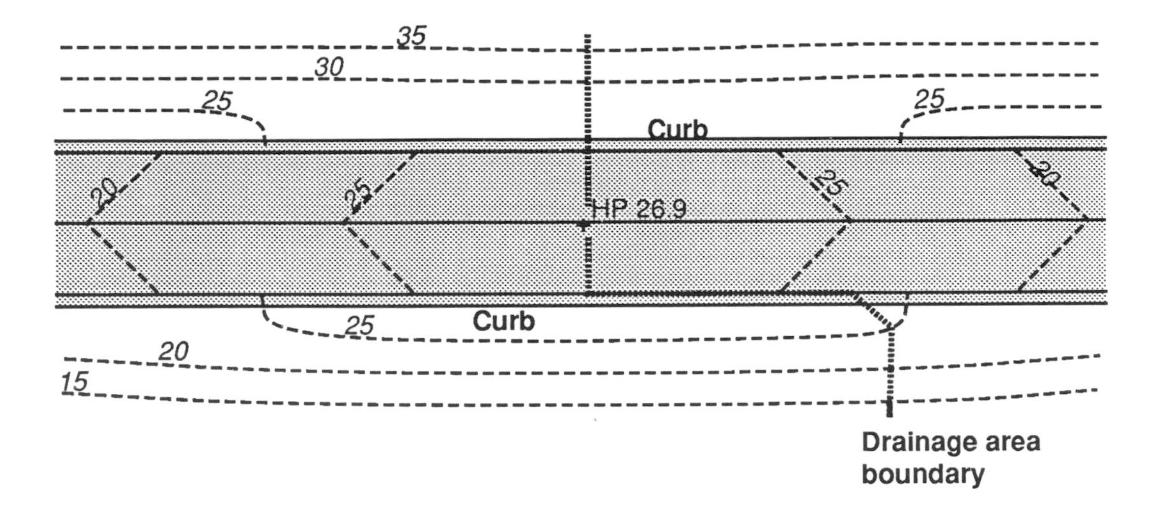

Similarly, approach a road from the lower side, following the contours as usual. Where the road has a curb, the curb on the downhill side of the road is likely to be the local topographic high line. So, go to the top of the curb and follow it uphill to the road's high point, where you will probably cross the road at a right angle to divide the flows of water into and out of your watershed. If the road has no curb, probably the road's crown is the topographic high line instead — so go to the crown and follow it as you would a curb or a roof ridgeline. After crossing the road at its high point let the topographic contours continue to guide you up the surrounding hillside.

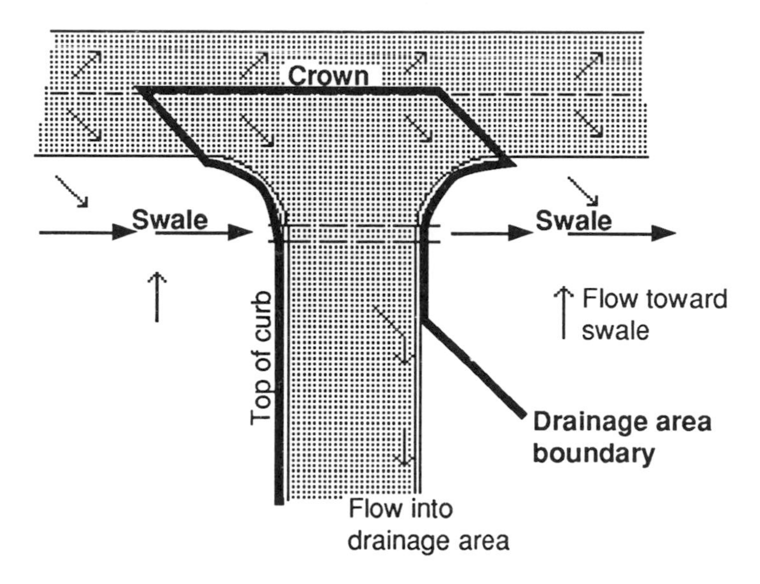

A tricky place on a site where we have worked is shown in the illustration. An existing road is located almost exactly on a topographic ridgeline. It has a central crown, no curbs, and a drainage swale. As part of a new residential development, a road with curbs is proposed to intersect with the existing road and to slope *down the hillside* into the development site. At the intersection, the crown of the existing road represents the boundary of the drainage area of the development's proposed storm detention basin. Runoff from the *existing* road flows onto the *proposed* road, where it is channeled into the development area by the curbs. Elsewhere along the existing road runoff continues to drain into the existing swale. A culvert carries the swale under the new road without any connection to the drainage of the new road.

If you have any doubts about whether you have drawn part of a drainage boundary correctly, you can test it by imagining a drop of water just to one side of the line starting to flow overland. On one side of the boundary the drop should run to the outlet; on the other side it flows away somewhere else. That is the definition of a drainage area's boundary.

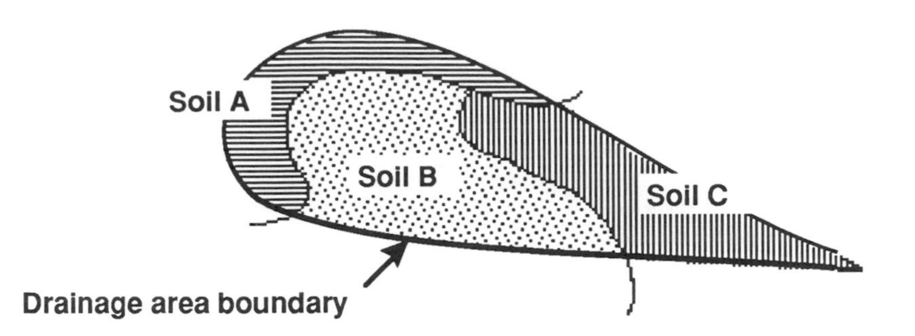

Having drawn the boundary correctly you are in a position to estimate a drainage area's size and to characterize its land use, soils, etc. The size in acres can be scaled from a site map.

For soil cover conditions, where there is only one soil type and one land use you can just look up the *CN* or *C* factor for that combination in the appropriate table.

Where there are two or more soil types or land uses you have to derive a composite *CN* or *C* factor. Tables such as those in Chapter 4 list some composite land uses. You can just read or interpolate those listings where appropriate.

But you have to derive a composite factor for a drainage area with, for instance, residential land use and three different soil types. Measure the area occupied by each soil type. You can derive a composite *CN* by weighting the *CN* of each combination of land use and soil according to how much of the total area it occupies. If we call the *CN*s of three soil-land use areas *A*, *B* and *C*, then the weighted average *CN* is given by:

$$\text{Average } CN = \frac{A(\text{Area } A) + B(\text{Area } B) + C(\text{Area } C)}{\text{Total drainage area}}$$

The resulting average applies to the drainage basin as a whole. Although the spatial distribution of *CN*s might be relevant to runoff from a drainage area, it is not taken into account in the commonly available rainfall-runoff models.

You can use the same type of weighted averaging for any number of combinations of soil, land use and any other necessary drainage area characteristics.

When comparing runoff before and after development such as for designing a storm detention basin you should understand that almost any characteristic of a drainage area could be altered by development, even though the area's outlet remains at the same location. Site-specific changes depend on land use, site design and preexisting site factors.

Of course, land use always changes. This involves changes in the percentage of impervious cover, vegetation and combined factors such as C or CN. Such changes can in turn alter the flow velocity and time of concentration.

Drainage area boundaries can change due to earthwork and the installation of roads and buildings. Roads have gutters or swales that divert runoff across hill slopes. The roofs of large industrial buildings can pitch runoff significantly in new directions. New boundaries mean that a drainage area's size and time of concentration may be altered. Soil type and slope may be altered if the new boundaries swing the drainage area into parts of the site with other soil types or topography.

Earthwork alone might change runoff substantially even when drainage area boundaries remain in essentially the same place. In a big public project such as a highway or an airport average slope can be changed by major earthwork. Major earthwork could also effectively change the soil type if it creates substantial areas of cut or fill which have different hydrologic characteristics from those of the preexisting surface soil. The local SCS soil survey might indicate variations of soil character with depth. In contrast, many residential developments are composed of a large number of relatively small individual units, often allowing site design to adapt to the preexisting topography more tightly without altering the average slope or soil of an entire drainage area.

In our exercise sites (Appendix A) we implied that drainage areas' size, slope, soil, etc., were essentially the same before and after development. That is a plausible assumption in the medium-density residential type of development used in the problem setup. In practice, you should check every relevant condition, beginning with drawing drainage area boundaries, both before and after development.

Appendix C

Showing a detention basin on a contour grading plan requires a few special features which do not occur in other places in grading plans.

The following discussion only illustrates the most simple and conventional forms of detention basins. Creative site design can call on many types of materials, plantings, shaping, and integration with other features of site developments to devise many variations on the basic vocabulary of components introduced here.

A dam is an earth fill or solid structure that spans from one side of a swale or valley to the other. One is necessary to make a basin on almost any site with moderate topography. (On an extremely flat site a basin might be made entirely by excavating.) An earth-fill dam has a more or less level top, a downstream side sloping outward, and an upstream side sloping back into the basin. A dam's proposed contours span from one side of the swale to the other.

A detention basin is empty storage volume created on the upstream side of a dam. It is represented by a combination of existing swale contours and proposed dam contours. These contours connect to each other to enclose the basin.

The illustration shows a very simple dry storm detention basin. Existing contours outline the natural swale where the basin is to be built. Proposed contours span from one side of the swale to the other to form a dam. The top of the dam is level and is at elevation 31. A culvert drains from the lowest point in the basin (elevation 13) to the existing swale on the downstream side of the dam (elevation 6). An emergency spillway is excavated in natural ground few feet lower than the dam.

During storms smaller than or equal to the design storm, the basin fills partly or completely with water, while the principal outlet controls outflow rate. When the storm inflow slows down and ceases, outflow continues through the culvert until the basin is completely drained and is "dry" again. When storms larger than the design storm occur, ponded water rises over the level of the emergency spillway, which passes the excess flows without eroding the dam.

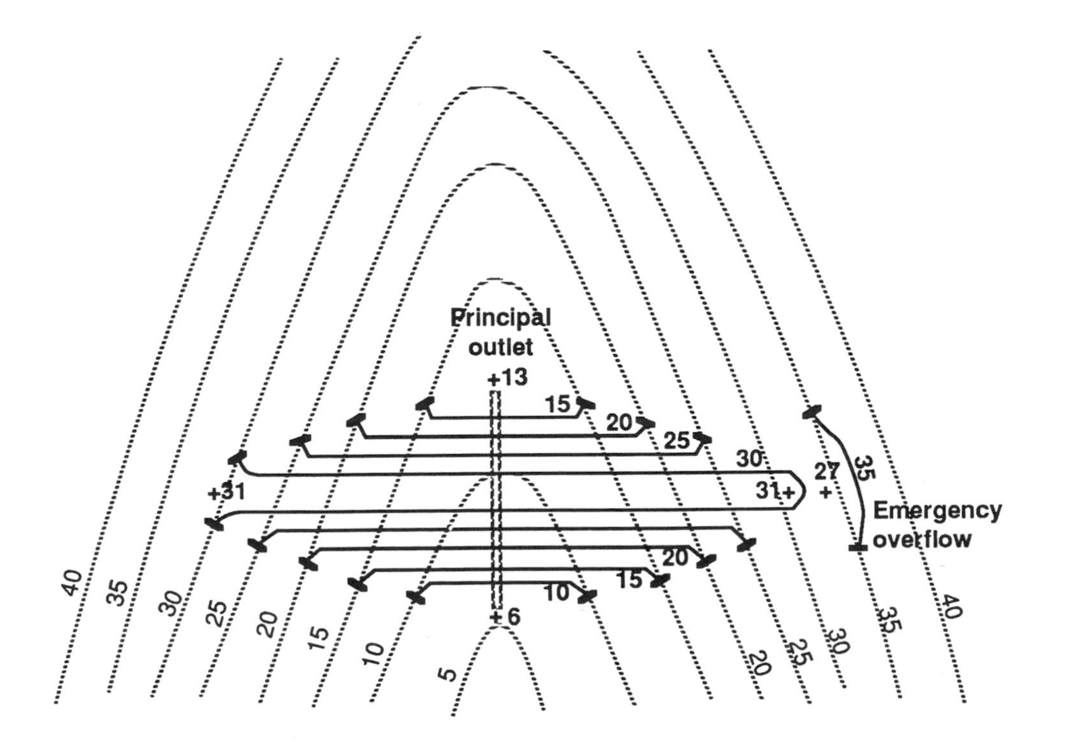

Emergency or "secondary" outlets are needed to protect dams. A secondary outlet is a nonerodible, nondamaging outlet to convey flow around the dam when a storm exceeds the design storm or the principal ("primary") outlet is blocked. It often takes the form of a vegetated channel excavated in stable cut, away from the fill of the dam. Its invert elevation is at the maximum elevation of the water surface during the design storm, or a half-foot or so higher as a safety buffer. The water surface may back up still higher during the very large storms for which an emergency outlet is designed. A separate secondary outlet may not be necessary where a principal outlet can convey very large storms, as could an appropriately constructed concrete weir.

"Freeboard" usually needs to be added to the height of a dam. Freeboard is the difference in elevation from the water surface associated with the emergency outlet, to the top of the dam. It is added to the dam as a safety buffer to take into account dam settlement, wave action in the basin, increased water elevation due to outlet blockage, etc. Freeboard of a couple of feet is common for many small on-site detention basins.

The illustration shows a "wet" detention basin where the surface of the permanent pool is established by a standpipe which acts as the principal outlet. A channel through the earth adjacent to the dam acts as the emergency outlet. Above the elevation of the principal outlet are a succession of ponding depths and freeboard.

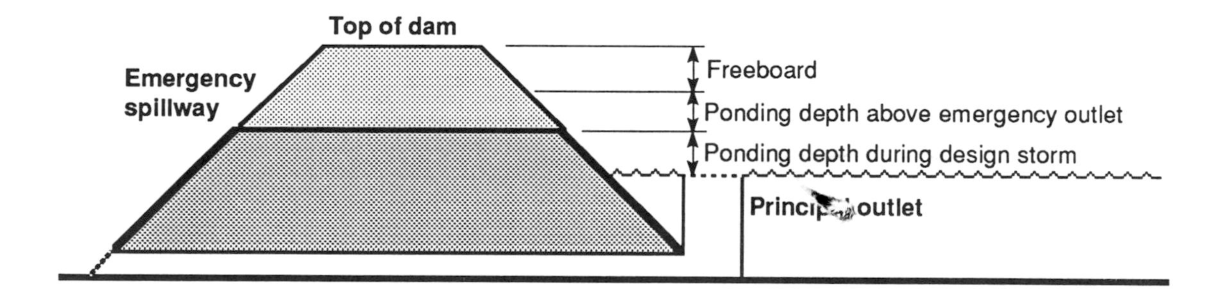

An energy dissipator is necessary where the velocity of water passing through a culvert or other outlet is capable of eroding the downstream channel. It often takes the form of an apron of riprap lining the swale around the downstream end of the outlet and extending downstream.

Appendix D

In any stormwater basin accumulated sediment and debris can reduce the storage capacity (volume) if they are in sufficient quantity, of the right type and without compensating design or management.

In an infiltration basin they can also reduce soil infiltration rates by clogging the soil surface, thereby prolonging ponding time and reducing the basin's capacity to handle further runoff.

Three types of sediment and debris reach stormwater basins, judging from experience on Long Island (Aronson and Seaburn, 1974):

1. Plant litter such as grass cuttings, leaves and twigs is most characteristic of occupied residential areas. Porous and friable, it does not reduce infiltration much, but it can clog small conveyances and narrow basin outlets from time to time.

2. *Automotive matter* and pavement material from roads and parking lots, such as asphalt, oil, rubber particles and pavement aggregate particles, is most characteristic of highways and commercial and industrial areas. It can greatly reduce infiltration rates, but is not likely to reduce storage capacities (volumes) or clog conveyances.

3. Mixtures of clay, silt and fine sand can come from any area not adequately protected against soil erosion. They could reduce infiltration rates, depending on particle size, sedimentation rates, and the vegetation and other characteristics of the receiving basin. They could reduce conveyance capacities or basin storage capacities if sedimentation is rapid relative to frequency of maintenance.

Think of urban sediment in two phases: 1) during construction, when soil is bare and very large quantities of sediment can be produced very quickly, and 2) after construction, when soil is expected to be stabilized by permanent vegetation.

Several ways to ameliorate the effects of sediment and debris have been suggested by research and experience:

During construction an unfilled infiltration basin can be used as a stormwater and sediment trap as long as the basin floor is not excavated lower than a foot above the final floor elevation. After the watershed is stabilized with structures and vegetation and sediment can no longer be generated, accumulated sediment should be removed and the basin excavated to final depth. Equipment and personnel should enter the basin floor only to the degree necessary to complete the excavation.

After constructuction, prevent sediment from any new earthmoving in the basin's watershed from entering the basin. Use a silt fence or other sediment trap at the sediment source or around the inlet rim until the source is eliminated.

Live vegetation appears to be effective in maintaining infiltration rates in open dry basins. Bermuda grass has repeatedly withstood long periods of submergence and drought without a major decline in soil cover. In California infiltration rates were substantially increased after Bermuda grass was established in a dense turf with a mulch of old grass. The grass grew up through accumulating silt deposits, forming a porous turf and preventing formation of an impermeable layer. Breaking up or loosening floors of infiltration basins is practiced on Long Island with the intent of raising soil infiltration rates, particularly when converting the use of the basins from sediment trapping to stormwater infiltration. However, the effect of scarification on basin performance has not been monitored. In California it was found that any mechanical disturbance of a soil that contains fine material actually reduced infiltration rates. In one case leveling the floor of a basin by moving the moist, fine-textured soil with heavy equipment greatly compacted the soil, reducing infiltration to only 8% of its original rate.

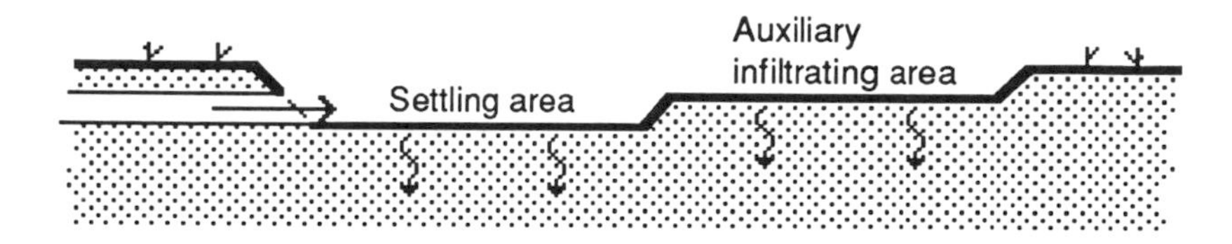

Sediment can be settled or filtered out before it reaches a basin. On Long Island the floors of some open infiltration basins have been built on two levels: in a lower level to which stormwater is delivered by pipes or swales, trash and sediment collect and some water may be infiltrated; an upper level infiltrates clean water that occasionally overflows the flooded lower level. Also on Long Island, the soil surfaces in a dozen or so infiltration basins draining heavily traveled highways and very large parking lots have become permanently clogged by inordinately large accumulations of automotive and pavement matter. The effort of continual rehabilitiation was too costly so these basins are now used only as ponds for settling out sediment and debris. The sediment-free overflow discharges to adjacent basins for infiltration. Presumably a similar dual-basin approach could be used for proposed basins, as well. In Arizona a vegetated (grassed) filtration strip over which water flows has been recommended for sensitive water harvesting applications.

Inspect basins in months when little or no standing water is expected according to the water balance, and at any other times if overflows are observed. A properly constructed underground basin can be inspected through the inlet like an ordinary catch basin. Pump or otherwise remove standing water and accumulated sediment when they are found, then wait for the next scheduled inspection.

Undertake the expense of major rehabilitations only after pumpage has been necessary at unacceptably frequent intervals, and has not been due to extraordinarily large storms or temporary sediment production in the watershed. Reconstruction should include brushing or scraping the floor and sides to remove caked silt, replacing clogged filter fabric, and cleaning or replacing stone fill.

Appendix E

The following procedure for estimating time of concentration is that specified for the *TR-55* version of the SCS method (U.S. Soil Conservation Service, 1986).

Three specific flow conditions are distinguished. In order of occurrence going down the watershed, they are sheet flow, shallow concentrated flow and open-channel flow. The hydraulic length is divided into segments, each segment having one of these conditions. Time of travel is found in each segment separately, then summed to find total time of concentration. The following pages describe SCS's procedure for finding time of travel in each of the segments. The total *Tc* should never be less than 0.1 hour (6 minutes).

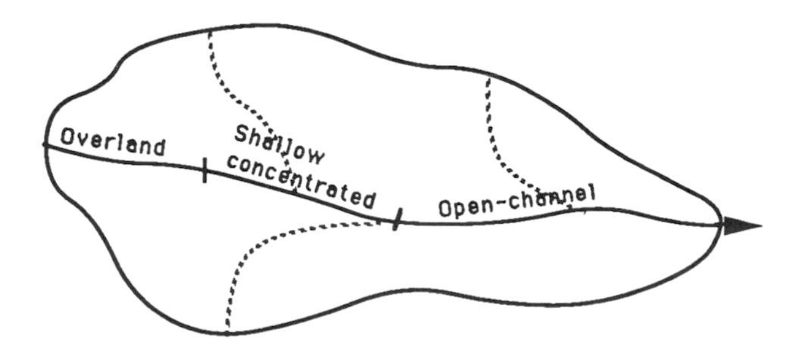

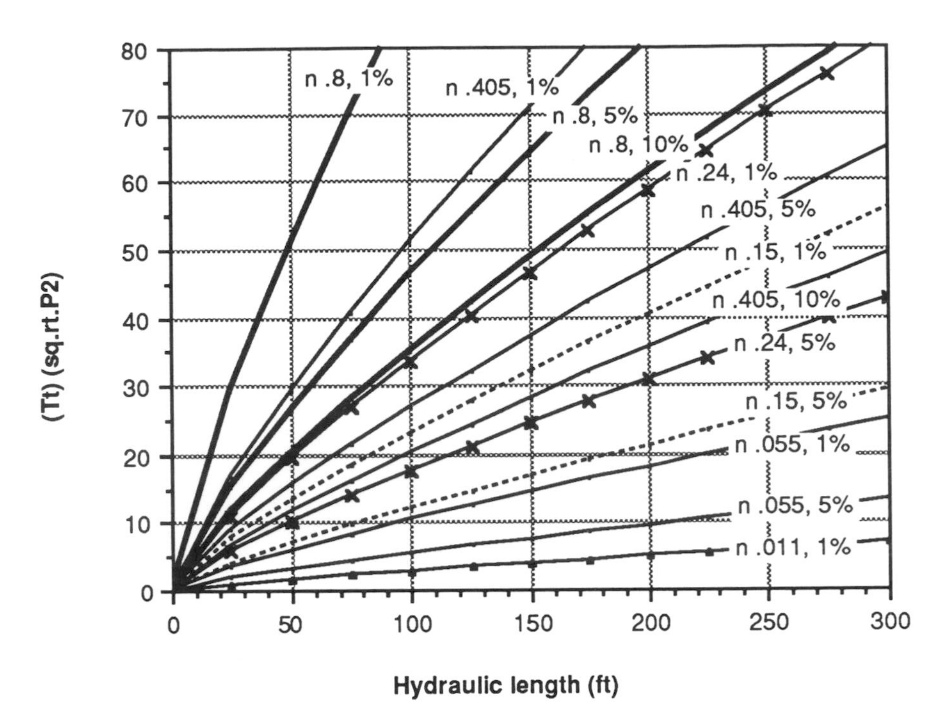

Chart was derived from Equation 3-3 of U.S. Soil Conservation Service, 1986, *Urban Hydrology for Small Watersheds*, Technical Release 55, second edition.

Sheet flow is flow over plane surfaces such as hillsides and parking lots. It usually occurs in the uppermost areas of watersheds. By definition, it never occurs for more than 300 feet. It is found specifically for conditions of a two-year storm, even though the two-year storm may not be the design storm for the facilities you are planning.

To find sheet flow time of travel, first find Manning's roughness coefficient *n* for the type of material the flow passes over, from the table on the next page. Find the length and slope along this part of the hydraulic length from a topographic map of your site. Enter the graph above from the bottom at your hydraulic length. Move up vertically to the line for your *n* and slope, then horizontally to read the time factor, (*Tt*)(sq.rt.*P2*), on the left-hand side.

Find two-year, 24-hour precipitation *P2* in inches from the map on page 35.

Time of travel *Tt* in minutes through this segment can then be found from,

$$Tt = \frac{(Tt)(\text{sq.rt.}P2)}{60\ (\text{square root of } P2)}$$

Manning Roughness Factors *n* for Sheet Flow

When selecting n consider cover only to a height of about 0.1 ft; this is the only part of the plant cover that will obstruct sheet flow.

n	*Surface description*
0.80	Woods with dense underbrush
0.41	Bermudagrass
0.40	Woods with light underbrush
0.24	Dense grasses such as weeping lovegrass, bluegrass, buffalo grass, blue grama grass, and native grass mixtures
0.17	Cultivated soils, >20% residue cover
0.15	Short grass prairie; cultivated with residue cover > 20%
0.13	Natural range
0.06	Cultivated soils, ≤20% residue cover
0.05	Fallow (no residue)
0.011	Smooth surfaces (concrete, asphalt, gravel, bare soil)

From Table 3-1 of U.S. Soil Conservation Service, 1986, *Technical Release 55, Second Edition*..

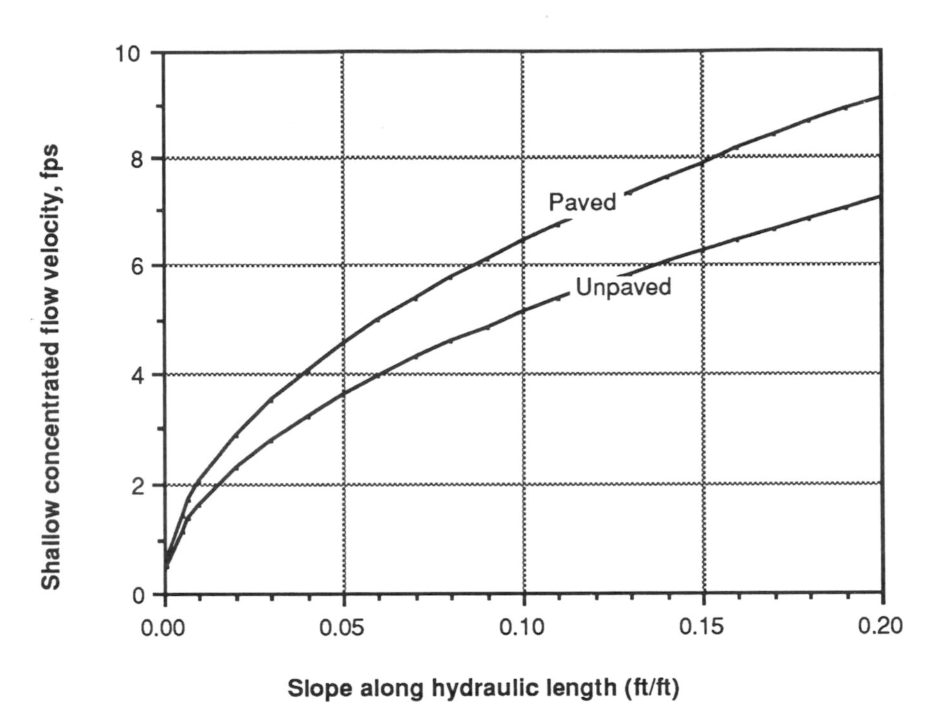

Chart was derived from equation for Figure 3-1 given on page F-1 of U.S. Soil Conservation Service, 1986, *Urban Hydrology for Small Watersheds*, Technical Release 55, second edition.

Shallow concentrated flow takes over from sheet flow at least within 300 feet of the top of the watershed, unless channel flow has started within this distance.

Find length, slope and cover material of this segment of the hydraulic length from a topographic map of your site.

Find velocity by entering the above graph at the bottom, at the appropriate slope. Move vertically up to the line for your surface material, either paved or unpaved, thence horizontally to the left to read the velocity of flow in feet per second.

Find time of travel in minutes through this segment from,

$$Tt = L \ / \ 60V$$

where,

Tt = travel time in minutes;
L = length of shallow concentrated flow's segment of hydraulic length, in feet;
V = velocity in feet per second.

Open-channel flow occurs in any pipes or definable swales and stream channels that occur along the hydraulic length. It takes over from sheet flow or shallow concentrated flow at any point where it first occurs going down the hydraulic length. Once a channel flow condition begins, the hydraulic length is likely to remain in channel flow for the remainder of the watershed.

Defined channels are assumed by SCS to begin at any point where surveyed cross-section information has been obtained, where channels are visible on aerial photographs, or where blue lines (indicating streams) appear on U.S. Geological Survey topographic maps.

Find length, slope, hydraulic radius and roughness factor *n* for this segment of the hydraulic length from a topographic map and the table on page 84.

Find velocity of open-channel flow by applying Manning's equation (page 83), with the assumption that the channel is full-flowing or bank full. The charts on pages 86 and 87 solve Manning's equation for velocity under a variety of hydraulic conditions.

Find time of travel through the channel from,

$$Tt = L / V$$

where,

Tt = travel time in minutes.
L = drainage area's hydraulic length in feet.
V = velocity in feet per minute.

Summary of Process
SCS Time of Concentration

Define flows:

1. Identify the hydraulic length of the watershed on a site map.

2. Divide the hydraulic length into sheet-flow, shallow concentrated flow and open-channel flow segments.

Sheet-flow time of travel:

3. Obtain length *L1* in feet and slope in ft/ft of sheet-flow segment from a site map. The segment must not exceed 300 feet in length.

4. Obtain roughness coefficient *n* (no units) from the table on page 221.

5. Obtain time factor (*Tt*)(sq.rt.*P2*) (no units) from the graph on page 220.

6. Obtain rainfall of a two-year, 24-hour storm *P2* in inches from the map on page 35.

7. Calculate time of travel *Tt*1 of segment in minutes from the equation,

$$Tt\,1 = \frac{(Tt)(\text{sq.rt.}P2)}{60\ (\text{square root of } P2)}$$

Shallow concentrated flow time of travel:

8. Obtain length *L2* in feet and slope in ft/ft of shallow concentrated flow segment from a site map.

9. Obtain velocity *V2* in fps of segment from the graph on page 222.

10. Calculate time of travel *Tt*2 of segment in minutes,

$$Tt2 = L\,/\,60V2$$

Open-channel flow time of travel:

11. Obtain length *L3* in feet, slope in ft/ft and hydraulic radius *R* of open-channel flow segment from a site map.

12. Obtain roughness *n* of segment from table on page 84.

13. Obtain velocity in fpm from the appropriate chart on page 86 or 87, or Manning's equation.

14. Calculate time of travel *Tt*3 of segment in minutes,

$$Tt3 = L\,/\,60V3$$

Total time of concentration:

15. Calculate time of concentration *Tc* in minutes by summing the times of travel in the three flow segments:

$$Tc = Tt1 + Tt2 + Tt3$$

Time of Concentration Exercise

1. Before Development

		Site 1	*Site 2*
Sheet-flow time of travel			
Sheet-flow length (300' max.) (from site map)	=	ft	ft
Sheet-flow slope (from site map)	=	ft/ft	ft/ft
Sheet-flow roughness coefficient n (from page 221)	=		
Time factor (Tt)(sq.rt.$P2$) (from page 220)	=		
Two-year, 24-hour rainfall $P2$ (from page 35)	=	in	in
Sheet-flow time of travel $Tt1$ = (Tt)(sq.rt.$P2$)/(sq.rt.$P2$)	=	min	min
Shallow concentrated flow time of travel			
Shallow concentrated flow length $L2$ (from site map)	=	ft	ft
Shallow concentrated flow slope (from site map)	=	ft/ft	ft/ft
Velocity of flow $V2$ (from page 222)	=	fpm	fpm
Segment time of travel $Tt2$ = $L / 60V2$	=	min	min
Open-channel flow time of travel			
Channel-flow length $L3$ (from site map)	=	ft	ft
Channel-flow slope (from site map)	=	ft/ft	ft/ft
Channnel-flow hydraulic radius R (from site map)	=	ft	ft
Channel-flow roughness n (from page 84)	=		
Channel-flow velocity $V3$ (from page 86 or 87)	=	fpm	fpm
Channel flow time of travel $Tt3$ = $L / 60V3$	=	min	min
Total time of concentration			
Time of concentration Tc = $Tt1 + Tt2 + Tt3$	=	min	min

Time of Concentration Exercise
2. After Development

		Site 1	*Site 2*
Sheet-flow time of travel			
Sheet-flow length (300' max.) (from site map)	=	ft	ft
Sheet-flow slope (from site map)	=	ft/ft	ft/ft
Sheet-flow roughness coefficient *n* (from page 221)	=		
Time factor (*Tt*)(sq.rt.*P*2) (from page 220)	=		
Two-year, 24-hour rainfall *P*2 (from page 35)	=	in	in
Sheet-flow time of travel *Tt*1 = (*Tt*)(sq.rt.*P*2)/(sq.rt.*P*2)	=	min	min
Shallow concentrated flow time of travel			
Shallow concentrated flow length *L*2 (from site map)	=	ft	ft
Shallow concentrated flow slope (from site map)	=	ft/ft	ft/ft
Velocity of flow *V*2 (from page 222)	=	fpm	fpm
Segment time of travel *Tt*2 = $L / 60V2$	=	min	min
Open-channel flow time of travel			
Channel-flow length *L*3 (from site map)	=	ft	ft
Channel-flow slope (from site map)	=	ft/ft	ft/ft
Channnel-flow hydraulic radius *R* (from site map)	=	ft	ft
Channel-flow roughness *n* (from page 84)	=		
Channel-flow velocity *V*3 (from page 86 or 87)	=	fpm	fpm
Channel flow time of travel *Tt*3 = $L / 60V3$	=	min	min
Total time of concentration			
Time of concentration *Tc* = $Tt1 + Tt2 + Tt3$	=	min	min

Appendix F

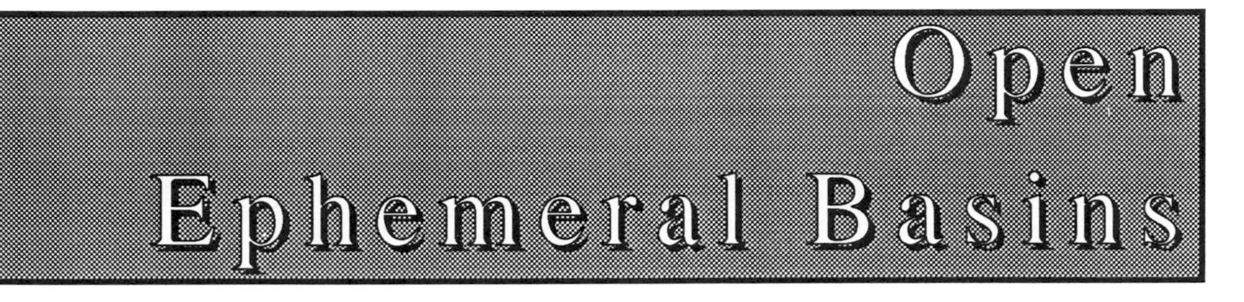

An open infiltration basin is more complex than an underground one. An open basin is exposed to direct precipitation, evaporation, and the overlap of the basin area and catchment area, none of which occurs in underground basins. In addition, the sloping sides of an open ephemeral basin cause the water surface area exposed to evaporation to change as the water level rises and falls, and create a complex relationship of water level to wet side area where infiltration occurs. In contrast, underground basins are presumed to have essentially vertical sides.

An ephemeral regime in an open basin can be evaluated using a monthly routing procedure similar in concept to that described for below-ground basins in Chapter 7, *Infiltration Design.* Refer to that chapter for background on ephemeral-regime routing and the Minnesota method of basin sizing, which incorporates both a basin's long-term water balance and its capacity to capture and hold the volume of runoff from a design storm.

This appendix supplements Chapter 7 by adding the additional flows to which open basins are uniquely subject.

Catchment-area overlap was introduced in the chapter on *Water Harvesting*. As the overlap area gets larger, inflow of runoff from the catchment is reduced. This effect can be taken into account by a loss-of-runoff term Ldr:

$$Ldr = Dr\ X$$

where,

Ldr = runoff depth lost due to catchment-area overlap, in ft.;
Dr = depth of runoff from the catchment area, in ft.; and
X = proportion of basin area that overlaps the basin's catchment area, decimal fraction.

Dr is estimated using the procedure discussed in the chapter on *Water Harvesting*. *X* can be roughly estimated from a site map. When multiplied by basin area, *Ldr* results in the volume of runoff lost *Ldr*.

Direct precipitation *P* and lake evaporation *El* are evaluated as depths of inflow. Monthly data for both are listed in *Water Harvesting*. When multiplied by basin area, precipitation depth yields the volume inflow *Pvol*; evaporation depth yields the volume outflow *Evol*.

The change in an open basin's storage from month to month, when the above flows are taken into account, is a volume,

$$\Delta \text{Storage} = Pvol - Evol - Ldr + Ro - Ivol$$

No basin fill is used in a open basin, so void ratio in fill does not need to be taken into account. The basin's capacity to hold water is equal to the full volume of the basin.

The safety factor *Fs* on infiltration rate need only take into account uncertainty about soil infiltration rate and possible compaction and sediment accumulation on the basin's floor. A value of approximately 0.5 is suggested. A lower value is not needed to protect roots of vegetation from saturated soil, since it is accepted that an ephemeral regime will result in prolonged standing water and soil saturation in any case.

A trial basin design can be evaluated, following the Minnesota method, by routing monthly storage and stage through it and computing the volume held in reserve for a design storm should it occur in any month. If the basin has adequate reserve capacity for the design storm every month, then it is big enough. If it has more than adequate capacity, then its size could be reduced. After revising the basin's design, whether to make it larger or smaller, redo the routing calculations to confirm the adequacy of the new basin design.

Development of the routing procedure discussed in this appendix is detailed in Bruce K. Ferguson, in press, Role of the Long-Term Water Balance in Management of Stormwater Infiltration, *Journal of Environmental Management*.

Summary of Process
Open Ephemeral Infiltration Basin

Trial basin:

1. Obtain storm runoff volume (*Qvol*) in ac.ft. after development, from your previous estimate on page 50.

2. Set initial trial basin dimensions in ft. for a basin with a trapezoidal cross-section, at least big enough to hold the design storm: floor length *Lf*, floor width *Wf*, total depth *Dt*, slide-slope gradient *G*, top length *Lt*, and top width *Wt*. Compute floor area in ac. $Af = LfWf/43{,}560$ and top area in ac. $At = LtWt/43{,}560$.

Monthly storage:

3. Identify the month with the greatest water deficit (the greatest excess of lake evaporation over precipitation) from the data used in water balance estimates on pages 68, 69, 71 or 72, and 76.

4. In the month with the greatest deficit, set initial basin stage *S*, storage *Vw*, and wet side area *As* wet all equal to zero.

5. Obtain monthly volume of inflowing runoff *Ro* in ac.ft. from your water-balance estimate on page 68, 69, 71 or 72.

6. Compute monthly volume of outflowing infiltration Ivol. With infiltration I in ft./mo. obtained from your estimate on page 76, a selected floor infiltration safety factor *Fs*, and a different hydraulic gradient *Ghf* from that on the sides *Ghs*, the infiltration volume in ac.ft./mo. is given by,

$$Ivol = (GhsAs\text{ wet} + FsGhfAf)I$$

7. Obtain average monthly precipitation *P* in feet/mo. from page 68,69, 71 or 72. Multiply by the basin's exposed surface area *At* in ac. to get acre-feet per month of direct precipitation inflow *Pvol*.

8. Compute area of upper water surface *Au* in ac.,

$$Au = (Wf + 2S/G)(Lf + 2S/G)$$

9. Obtain average monthly lake evaporation *El* in feet/mo. from page 76. Multiply by the basin's upper water surface area *Au* in ac. to get acre-feet per month of evaporation outflow *Evol*.

10. Compute volume of water storage *Vw* in ac.ft. at beginning of the following month,

$$Vw = \text{Previous } Vw + Pvol - Evol - Ldr + Ro - Ivol$$

11. Compute stage *S* in ft. at the beginning of the following month,

$$S = 2Vw / (Af + Au)$$

12. Compute the wet side area *As* wet in acres at the beginning of the following month,

$$As\text{ wet} = (S^2 + S^2/G^2)^{1/2}(Lf + Wf + 4S/G)/21{,}780$$

13. Compute the unfilled volume held in reserve for the design storm in ac.ft.,

$$\text{Reserve capacity} = (Dt - S)(At + Au)/2$$

14. Repeat the routing calculations until the stage at the end of each month is the same as that in the same month of the previous year.

Evaluation:

15. Compare the basin's minimum unfilled volume to the reserve capacity needed to hold the volume of the design storm.

16. If necessary, adjust the basin's dimensions and revise the routing calculations accordingly.

Open-Basin Exercise

Discussion of Results

The overall water balance of your development site during an average year can be evaluated using the data from your water-balance estimate and basin routing.

Precipitation is equal to the unaltered precipitation that fell on the site.

Surface runoff from the site is zero, if your basin is sized to capture all inflows during an average year.

Evaporation from the watershed upstream of the basin is equal to the difference between site precipitation and runoff entering the basin, unless the watershed is in an upland situation where infiltrated water could pass through the groundwater clear under the basin before emerging to surface streams. In such an upland situation, some portion of the site precipitation remaining after runoff would be allocated to infiltration; the magnitude of that portion could vary widely depending on site-specific climatic and geologic circumstances. The monthly evaporation from the basin is estimated as described on the previous page. The total evaporation from the site is that from the watershed upstream from the basin, plus that from the basin.

Total infiltration from the site is precipitation minus total evaporation, as long as surface runoff from the site is equal to zero.

The overall water balance should then resolve into the form,

Precipitation = Runoff + Infiltration + Evaporation.

Appendix G

We think that there are real things you can do to make stormwater facilities safer than they might otherwise be.

We have taken informal surveys at meetings in many parts of the country by asking designers to describe stormwater facilities they have seen where persons have gotten hurt.

Large quantities of fast-moving water, particularly constrictions of flows at the mouths of culverts, are involved in about two thirds to three quarters of the stories that designers tell.

An example is a stream in State College, Pennsylvania that drains a residential subdivision and then passes through a public park. Shortly after a large storm a four year old girl fell into the stream in the park and was trapped against a trash rack at the entrance to a culvert. The trash rack was properly sloped at a 45 degree angle, which in theory is supposed to allow entering water to push objects up and out of the way. But the little girl's body wasn't shaped right — the water had nothing for the water to get behind and push up on. In fact there was so much water entering the culvert that it was pushing *down* on her as it went through the trash rack. Four men couldn't pull her away and she was killed.

Another example is a town-owned dry detention basin in East Brunswick, New Jersey, located between two subdivisions. The basin's 21-inch culvert outlet discharges into a three-mile-long series of storm sewers. During a flood two 15-year-old boys watched the water rise above what was normally a weedy field to become a lake the size of a football field, 6 to 10 feet deep. Early in the afternoon a neighbor saw them throwing stones into the violent vortex over the outlet, and warned them away. But later in the evening, they were both missing from their homes. The dog that had accompanied them was found bound to a tree near the basin. The next morning searchers found their bodies, one a mile downstream in a 40-inch section of storm sewer, the other two miles farther down, where the storm sewers discharged into a creek. Since they were both fully clothed, possibly one of them fell in while watching the whirlpool, and the other also fell in while trying to help. The suction of the vortex and the enclosure of the pipes allowed them no escape once they slipped on the wet, grassy, steep-sided dam.

Large constricted conveyances are challenged by this kind of tragedy. Where large conveyances are necessary, keep them as open and freely flowing as possible, keep the flow broad, shallow and slow, and avoid sudden constrictions. At detention basin outlets, prefer broad, open, rectangular weirs to anything involving the mouths of pipes.

Still water in basins or ponds is involved in the other quarter to third of the stories that designers tell at our meetings. In essentially all of the still-water cases we have heard there have been steep, slippery side slopes of clay or wet grass, deep water and nothing to hold onto. If you were going to build a *trap* for people to fall into and drown, that is the way you would design it. There is nothing in the nature of still water that says it has to be surrounded by that type of environment.

To make still water safe take each of those designed-in hazards and do the opposite. You should make it unlikely that persons will fall in and, if they do fall in, able to pull themselves out again and able to be accessed for rescue. A basin should be visible, open and accessible, so that persons can appreciate any hazard and guard themselves, and so that rescue efforts, if necessary, are not slowed down. Mark the approaches to water with changes in ground cover, changes in grade, or unique features such as plantings, benches or terraces. Make basin side slopes gentle. Make the ground surface (both around the pond and on the bottom) rough, using pavements or ground covers. Make the water shallow. Break up water with islands, boulders and mounds; put solid objects for holding onto in the flood area, such as large plants or rocks. At outlets avoid sudden constrictions and high velocities.

The following two court cases (Kozlowski, 1985) illustrate some of the factors that have been considered in deciding liability for safety hazards.

In Illinois (Cope v. Doe, 1984) a seven year old boy drowned in a "wet" detention basin in an apartment complex. The Illinois court found the developer *not* liable for the boy's death. The pond was located 100 yards from the apartment buildings. The pond was clearly visible and was considered an "open and obvious danger" which the Illinois court believed children of age to be allowed at large should be expected to appreciate and avoid. On the day of the accident the pond was one-third covered with ice; open water could be easily seen. Children who had grown up in Illinois for seven years were expected to understand the hazard of thin ice. Nevertheless the boy went upon the ice surface and fell through while kicking pieces of wood. The apartment complex had been designed to attract families with young children. It included a children's playground, a swimming pool, tennis courts, and other recreational features. No precautions were taken to prevent children from going near the pond other than the manager's warning parents to keep children away from it. The primary purpose of the pond was flood control; although children were not prohibited from going near it they were not invited to it for swimming or other recreational purposes as they were to the swimming pool.

In Louisiana (Guillot v. Fisherman's Paradise Inc., 1983) a two year old boy drowned in a marina's sewage oxidation pond near weekend houses. The Louisiana court found the marina's developer *liable* for the child's death. The pond was located 100 yards from the house where the parents were staying but was completely unmarked; none of the adults were aware of its presence. There was no border around the pond except a three foot high earthen embankment; a partially completed wooden fence was only for aesthetic purposes. The pond had very steep side slopes and a level bottom about three feet deep. The adults had warned the child about the large nearby lake, which was the only water hazard of which they were aware. The child had never been allowed to go to the lake except in the company of adults, and was considered an obedient child. The child had wandered away from the house yard; within 10 minutes the adults noticed his absence and began to search for him. The Louisiana court did not expect parents ordinarily to keep watch over a young child *every* minute, or to keep a child in chains to prevent him from *ever* wandering. The boy had apparently wandered into the pond, which was covered with light green algae which made the presence of water less apparent, and contained some pieces of brightly colored litter which may have been attractive to a young child. Almost an hour had passed before the parents discovered the child's body floating face down in the pond.

Appendix H

Making a backup copy of the program disk before you begin using it could prevent you from losing its contents during computer operations.

The model assists students in estimating storm runoff according to the SCS method, and routing storm flows through conveyances and detention basins. By using the model to go rapidly through a number of calculations a student can learn the effects of different estimation assumptions and design alternatives. Although the model generates realistic results consistent with the original SCS method, it is designed for classroom use only and has not been calibrated for accurate application to site-specific design in any particular region. In practice, your professional judgment should never be replaced by a programmed model or its results.

This appendix deals only with the operation of the model on a computer. To use the model for an intelligent purpose you need to know some stormwater hydrology first. Refer to chapters in the text as you need to. Runoff estimation is discussed in Chapter 2. Channels are discussed in Chapter 4. Reservoirs such as storm detention basins and infiltration basins are discussed in Chapters 5 and 7.

The model works on an IBM or compatible computer. Only 256K of computer memory is required to run the basic program, but up to 640K may be necessary to make full use of the model's capability to handle numerous subwatersheds, channels and reservoirs. No graphics "card," plotter or other special graphics hardware is necessary.

The disk contains five separately named modules or programs, which the model uses sequentially for estimating runoff and routing downstream flows. The files are:

STARTUP — for you to set default drives and initialize local rainfall and other data;
INTRO — presents informational screens introducing the model;
DATA — for you to enter, store and organize site data;
HYDROS — performs the hydrologic calculations (stands for "hydrology-subwatersheds");
HYI — puts hydrographs from all subwatersheds on equal time basis for routing (stands for "hydrograph interpolation"); and
ROUTE — routes hydrographs through channels and reservoirs.

You need a blank formatted disk on which to store data relating to your own calculations, in addition to the program disk. As an alternative, you can have a hard disk, on which either or both of the model and the your data can be stored. Using a hard disk for the model and data storage will greatly increase the speed of calculations and model output. The six model files are not copy protected, and can be transferred to your hard disk using an ordinary copy command.

The model is menu-driven and interactive. By displaying information and questions on your screen the model tells you when to enter each piece of data necessary for your calculation. All your "yes" and "no" answers can be entered by pressing *Y* or *N* (either capital or lower case letters will work). After you enter any answer or data, always press the return key.

You should be able to operate the model without mishaps if you take a careful look at the on-screen instructions and choices before pressing any buttons, and refer to this appendix when necessary. The disk accompanies this book at a very low price in order to be widely accessible to students. Consequently neither the authors nor the publisher can afford to offer support for users either by telephone or by mail.

Routing Methods

As a given volume of water moves downstream in a flood wave through a network of subwatersheds, channels and reservoirs it is attenuated by temporary storage in each network member. On a hydrograph, its time base is lengthened and its peak flow is reduced. Given the flow at an upstream point, routing may be used to compute the flow at a downstream point. A routing calculation works its way downstream, combining inflows at the top of each member and routing flows to the bottom.

The SCS runoff estimation method generates subwatershed hydrographs by a kind of overland flow routing. Since the SCS method does not specify how to continue to route flows downstream through channels and reservoirs, this model supplements the SCS method with widely accepted channel and reservoir routing methods. Detailed discussion of each of these methods can be found in textbooks of hydrologic modeling. As for any programmed computer solutions, studying the methods with pencil and paper will help to justify any faith placed in the computer output.

Channels are routed using the Muskingum method. It estimates outflow in each time step from the lower end of a channel as a linear function of inflow and outflow during the previous time step. During the *n*th time increment,

$$O_n = C_o I_n + C_1 I_{n-1} + C_2 O_{n-1}$$

where,

I = inflow rate;
O = outflow rate;
C_o = $(-Kx + 0.5t)/(K - Kx + 0.5t)$
C_1 = $(Kx + 0.5t)/(K - Kx + 0.5t)$
C_2 = $(K - Kx - 0.5t)/(K - Kx + 0.5t)$
x = weighting factor between 0 and 0.5, typically 0.2-0.3 for natural channels;
K = storage time constant approximated by reach travel time.

Reservoirs are routed using the Puls method. The basic Puls equation is derived from the basic reservoir balance. During the *n*th time increment,

$$I_n + I_{n+1} + (2S_n/t - O_n) = 2S_{n+1}/t + O_n + 1$$

where,

S = volume of storage; and
t = time increment.

STARTUP Module

The following instructions assume you are loading the model in drive A and storing data on drive B. If you are using other drives, insert the appropriate drive designation in these instructions.

Turn on your computer and boot up from your DOS system. When *A*> appears on the screen, insert the program disk in the A drive.

If this is the first time you have used the disk, type STARTUP and press *Return* to begin the model. You need to run the STARTUP module to set default drives on your computer system and to enter local rainfall and other data for your local area and company. These data will be stored with the model on the disk, and can be used any number times without rerunning the STARTUP module. You don't need to run this module again until you want to change any of the data you first entered. To bypass the STARTUP module when you use the model later, respond to the *A*> by typing INTRO instead of STARTUP.

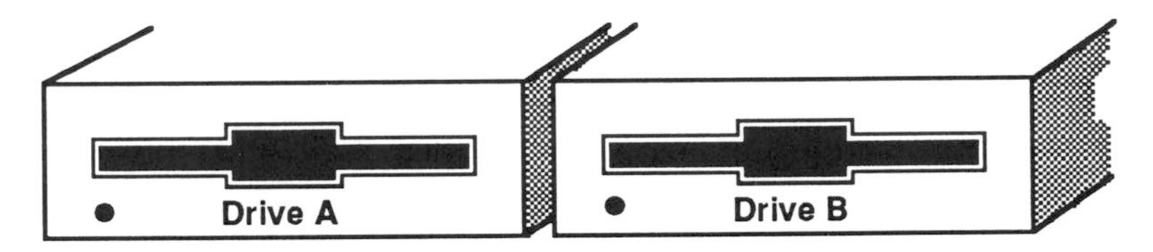

The STARTUP menu will come onto your screen. It presents nine options, which you can choose in any order. Select an option from the menu by entering the number of your choice.

Each of the first five options allows you to enter new data, or the review and edit data previously entered. The first time you run the model, be sure to run each of the first five options in the menu so that the model will have all the initial data it needs. If you select one of these options, the computer screen will show you what data need to be entered, and will display previously entered data if any. To enter or revise data, simply use the arrow keys to move the cursor to the necessary position, type in the value you desire and press *Return*. If other instructions are needed they will be given on the computer screen.

You should be prepared to enter data for each of the first five options:

1. Rainfall Volume: Enter data for the 2, 5, 10, 25, 50 and 100 year storms. The rainfall should be 24-hour rainfall in inches as used in the SCS method. Maps for three of these storms are on pages 35 and 36. You can also use other sources of data listed on page 35.

2. Soil Moisture: Choose one of four options for soil moisture conditions: dry, average, wet or changing. The model uses the condition you choose to adjust the Curve Number. If you specify "changing" conditions, the model keeps track of rainfall during the storm and changes the Curve Number as the rainfall accumulates. It bases its values on linear interpolation between the following successive limits:

Rainfall < 1.4 inches:	Curve Number for dry conditions;
Rainfall 1.4 to 2.1 inches:	Curve Number for average conditions;
Rainfall > 2.1 inches:	Curve Number for wet conditions.

3. Channel Shape: Enter the Muskingum routing coefficients for seven different channel shapes. Selecting Muskingum routing coefficients requires experience and sometimes measurements of velocity and flow along channel reaches of interest. Selection precedures are described in most hydraulic textbooks. The following table of values is presented as a starting point for selection of coefficients and should not be assumed correct for any specific application:

Channel shape	Coefficient
Rectangular	.33
Vee	.39
Flat trapezoidal	.20
Moderate trapezoidal	.25
Steep trapezoidal	.30
Earth-lined	.40
Concrete-lined	.45

4. Company Name: The name you enter here will be printed on all output from the operation of the disk, to identify the output as yours.

5. Default Drive: Select the default drives where you want the modal and data to be stored. It can be either a floppy disk or a hard disk. With this information on the model disk, you do not need to specify the correct drive every time you enter a new data file name.

The last four options are for saving or not saving the newly entered STARTUP data, quitting the STARTUP module, and going either back to DOS or continuing with the model by loading the DATA module. Select the one you want:

6. Save New Data and Load Data Module;
7. Don't Save New Data and Load Data Module;
8. Save New Data and Quit to DOS;
9. Don't Save New Data and Quit to DOS.

If you select number 6 or 7 the module for entering data about site-specific watersheds will be loaded into the computer's memory.

Entering Subwatershed Data

The DATA module boots up when you exit from STARTUP using the appropriate STARTUP option. You can also enter it directly, bypassing the STARTUP module, by typing INTRO instead of STARTUP when you first boot up the model. Some introductory informational screens will appear; you can speed up transition to the next step by pressing *Return* at each screen (if you don't press *Return*, the screens will be replaced automatically in a few seconds).

The program will ask whether you want to use data you have previously stored on your data disk. Answer *Y* or *N* and press the return key. If you answer *Y* you must then identify the job name under which your data are stored on your data disk. If you answer *N* the program will ask for a new job name.

Your job name is the label under which your data will be filed and stored on your data disk. You can use up to eight letters and numbers for such a name. Use a different name for each new data set in order to avoid overwriting and losing your previous data. One way to prevent loss of valuable data is by periodically making a backup copy of your data disk.

Before entering a new file you should have sketched out a correct network diagram. A discussion of networks diagrams begins on page 252.

All the necessary data for applying the SCS method to each subwatershed in your site will be systematically requested by the computer. You can specify data for portions of a large drainage area ("sub-watersheds"), or you can enter a single total watershed by referring to it as a single "subwatershed." You should be prepared to enter the following data about each of your subwatersheds (enter only numbers, not units):

- drainage area in acres;
- watershed's 85% elevation in feet (the elevation at the point on the hydraulic length 85% of the horizontal distance from the area's outlet);
- watershed's 10% elevation in feet (the elevation at the point on the hydraulic length 10% of the horizontal distance from the area's outlet);

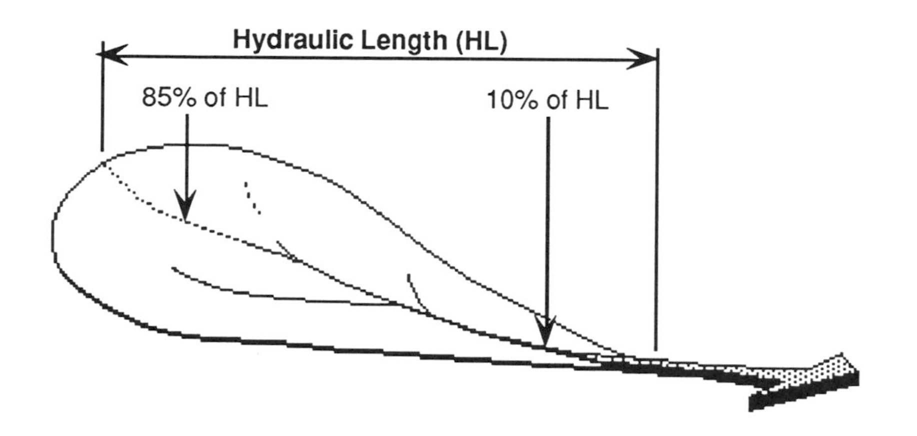

- watershed's total hydraulic length in feet or miles;
- watershed's impervious area in acres or percent;
- channelized length in feet or percent (the distance along the watershed's hydraulic length that has been modified in any way that could significantly increase velocity, such as paving the channel, cleaning and straightening the channel, installing storm sewers, etc.);
- curve number as defined in the SCS method; you can enter a curve number, or press *Return* and a curve number table will appear on the screen so you can make a selection.

After you enter the data for *each* subwatershed the program will give you an opportunity to check your data. If you find that there is an error in your data you can correct it by following the instructions on the screen.

After you have entered the data for *all* your subwatersheds just press *Return* at the beginning of the next subwatershed data request, and the program will know that your data entry is complete. You can enter data for up to 50 subwatersheds. If you specify a very large number of subwatersheds, channels and reservoirs, you may exceed the memory capacity of your computer.

The computer will ask if you want to run the routing program, as a transition to the routing module.

Answer *N* if you want only an estimate of runoff from a single subwatershed or several subwatersheds without routing through downstream channels or reservoirs. The program will then automatically store all the subwatershed data you entered on your data disk, under the job name you specified earlier. It will also ask whether you want to make a "map" (a printout in matrix form) of your input data and whether you want to make any changes in your data. The HYDROS module will be automatically loaded up; it will compute peak rate of storm flow and a complete storm hydrograph, after asking some preparatory questions. Detailed discussion of the HYDROS module begins on page 248. You should be prepared to enter the recurrence interval of your design storm, and the rainfall distribution type in your region (see the map on page 44). You can then run the data input module again, that is, you can enter or load from your disk a new set of watershed data.

Answer *Y* if you want to connect the flows of several subwatersheds together in a drainage network, or if you want to evaluate the effect of detention or infiltration basins.

Entering Drainage Network Data

When you answer *Y* to the computer's offer to begin routing, the HYDROS module is automatically loaded up, and will begin to ask for necessary data. When working on large, complex networks, sketching a network diagram on paper before entering data into your computer could make data entry a lot smoother. See the section on network diagrams later in this appendix.

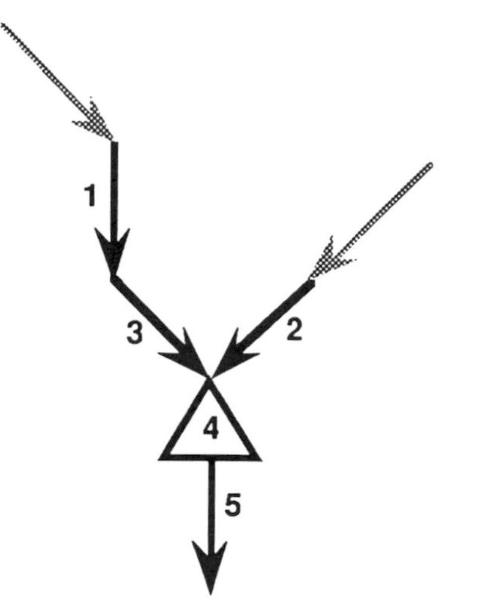

About your overall network, you should be prepared to enter the following data:

- Units of measuring storage, either cubic feet or acre-feet.
- Total number of members in the network; this is equal to the highest number on your network diagram (you can have up to 100).
- The number of reservoirs (detention or infiltration basins) in the network (you can have up to 20).

Data about channels is then entered. For each channel, separate the required data items by commas, such as 1,6,400,4. But do not put commas within one data item; for example enter the single number "ten thousand" as 10000 not 10,000. End each channel segment by pressing *Return.* For each channel in your network, be prepared to enter the following data:

- Channel number in the watershed's network diagram;
- Channel length in feet;
- Average velocity of flow in fps;
- Channel shape. From the eight descriptions listed on the screen, select the alternative that seems closest to actual cross-sectional shape. The first five choices are for irregular, coarsely vegetated natural channels:
    1. rectangular (natural, vertical sides, any width);
    2. vee (natural, triangular, with any side slope);
    3. trapezoidal (natural), with relatively flat side slopes;
    4. trapezoidal (natural), with moderate side slopes;
    5. trapezoidal (natural), with steep side slopes;
    6. smoothly aligned, earth-lined channel, with any shape;
    7. concrete-lined, with any shape; and
    8. "dummy" channel (no effect on routing of storm flow), used only for connecting subwatershed hydrographs to the network; for a dummy channel you can just enter channel number 8 and then press *Return.*

Data about reservoirs (lakes, ponds, detention and infiltration basins) are then entered. For each reservoir be prepared to enter:

- the number assigned to it on the network;
- the number of values you want to use in its "rating curve" (stage-storage-discharge curve); you can have up to 10.

Each rating curve value consists of a stage (height in feet above the outlet or above a base level), the discharge from the reservoir at that stage (in cfs) and the volume of storage in the reservoir at that stage (in cubic feet or acre-feet). The first discharge must always be zero. For a reservoir that will be empty at the start of the design storm, enter 0,0,0 as the first rating curve value. Discharge and storage must increase at each increase in stage. If you want discharge or storage to remain constant between two stages, enter a very small increase, such as from 50 cfs to 50.00001 cfs, to prevent errors resulting from division by zero. The program will interpolate linearly between all values given.

An infiltration basin is a special kind of reservoir. You might wish to route a storm on a site where infiltration is proposed in order to confirm that peak flows will be controlled, or to find the quantity of runoff remaining from the portions of the site that do not drain into infiltration basins. Here is an example of a rating curve for an infiltration basin:

Stage (ft.)	Discharge (cfs)	Storage (ac.ft.)
1.7	0.0	1.3
6.0	0.1	4.4
6.1	999999	4.4001

1.7 feet is the maximum depth to which average monthly background flows fill this basin; 1.3 ac.ft. is the volume of water in the basin at that stage. Consistent with the Minnesota method for sizing infiltration basins, 1.7 feet is taken as the stage at the beginning of the design storm. 6.0 ft. is the stage at the top of the basin. Although there is actually no discharge from the basin at this stage, the model cannot interpolate between two discharges of zero, so 0.1 is entered as a positive number which closely approximates zero. 4.4 ac.ft. is the total volume of water that the basin can hold (this is a stone-filled basin, and the total volume including stone fill is not shown). At the first stage higher than six feet, the basin overflows and does not restrain flow at all, so a very high potential rate of discharge is shown. At this stage there is no increase in storage over that at the top of the basin, but a very small increase must be shown in order to prevent division by zero. Infiltration basins will be listed with other "detention" basins in the model output.

Data about network connections are then entered. Refer to your network diagram (a discussion network of diagrams is at the end of this appendix). Each channel or reservoir can have up to five other network members flowing directly into it. The model automatically assigns the hydrograph from each subwatershed to go into the member having the same number, and will skip those members when asking you to enter further data about network connections. The order of entry does not affect the analysis.

Data Editing

All entered data about subwatersheds and the drainage network will be stored on your data disk under the job name you had specified.

You can make a printout or display of all data in the form of a table (a "data map") by answering *Y* when the program asks whether you want one. Using the output, you can check all your data.

You can revise any data necessary. Although editing a few items in a file is very easy, as a general rule, if you are going to make major revisions to your data such as by inserting or deleting many members, it is easier just to create a whole new file and start again. To edit data, when the program asks whether you want to make any changes answer *Y* and press *Return*. The model will then display all data in a sequence of screens. Upon reviewing each screen you have three choices:

- Press *Escape* to edit data on that screen;
- Press *Page Up* to see the previous page of data;
- Press any other key to see the next page of data.

If you press *Escape* to edit data, you then have three further choices:

• To revise previously entered data, use the arrow keys to move the cursor to the data item you wish to change. Type in the new value and press *Return* to replace the old value with the new one. Then move the cursor to any other item you wish to change, and continue replacing values.

• Insert or delete entire channels, reservoirs or other data items by moving the cursor to the data for the type of feature you want to insert or delete, such as a channel segment or reservoir. For either inserting or deleting, you have two options:

<u>If inserting:</u>

A. If you had positioned a dummy channel where you now want to insert a member, just insert the new member and delete the dummy channel. In this way, the numbering of all other members will not change.

B. Insert the new member and the model will change the numbers of all downstream members.

<u>If deleting:</u>

A. Delete the member you select and insert a dummy member with the member number. The numbering of all other members will not change.

B. Delete the member you select and let the model renumber downstream members.

Then move the cursor to any other type of feature you wish to insert or delete, and continue editing. After inserting or deleting members, completing all other data changes, and saving the new data set, be sure to check the network diagram table to be sure the correct members flow into each member. Edit this table if necessary.

• Press *Escape* to leave the edit mode for this page. Then you can proceed to other pages of data, where you might wish to enter the edit mode again. You must leave the edit mode before you can change pages.

A printout of your new data will be offered after you have reviewed all data in the file. Answering *Y* at this point will provide a final check of the data and a record for later reference. After printing out, the model will give you another opportunity to make changes, which you can take if necessary. If you answer *N*, the program will automatically store your corrected data on your data disk. You might also want to make a new backup copy of your data disk.

A new file name for saving the edited data will then be offered. If you use the old file name, it will overwrite all the old data. If you choose a new file name, you will save the new data in a file separate from the old data.

Storm Event Simulation
(The HYDROS Module)

The module for doing stormwater calculations will automatically be loaded into the computer's memory, as soon as you have entered drainage network data and are done with any editing.

You should be prepared to respond to the model's queries about the following options:

- The time of concentration estimated by the model is based on watershed slope. If you wish to use a different time of concentration, answer *Y* when the model offers. You will later be able to enter your *Tc* when the model needs it for runoff calculations. If you answer *N* the model will proceed with its own estimate.
- Storm rainfall distribution types are offered on the screen, four for conventional SCS storm types, and two developed specially for the state of Florida. See the map on page 44 to see which SCS rainfall region your site is located in.
- You have the option to print the hydrologic results either on the printer for a hard copy, or on the screen. Selecting screen display will allow you to see only the summary results (peak inflow, outflow, stage and storage). Outputting to the printer is the only way to get complete hydrograph data from all the intermediate calculation steps.
- The storm recurrence interval that you select will tell the model to use a rainfall volume that you specified in the STARTUP module. If you want to run all six possible design storms, enter *1* and press *Return*. If you choose to run all six storms, output will be sent automatically to the printer. Make sure your printer has enough paper.

The model will then perform all runoff calculations using the SCS method and store the resulting subwatershed hydrographs.

If the capacity of a reservoir is exceeded, the computer will sound a warning. The model is forced to cease calculations since it did not have data telling the reservoir what to do with such large flows. You can edit its rating curve to enlarge the reservoir, or revise your site plan to reduce runoff reaching that reservoir. To avoid loss of time from terminated calculations, you can initially enter reservoir data indicating excessively large storage and discharge capacity (usually associated with a taller dam), then decide on an exact, lower dam height after the calculations tell you how high the water actually rises in the basin during the storm.

The printed routing output consists of the following:

- for each subwatershed, a complete hydrograph, listing time, ordinate (flow rate in cfs), and volume, along with watershed calculation data such as time of concentration.
- for each channel, list of inflowing members and a hydrograph table showing *T* (time in hours), *I* (incoming flow in cfs), and *O* (outgoing flow in cfs);
- for each reservoir, a list of inflowing member and a table showing *T*, *I*, *O*, *V* (volume of storage in units you have chosen), and *H* (height of water in feet).

The summary output, whether printed or on the screen, consists of:

- For each subwatershed, the peak rate of runoff in cfs;
- For each channel, the peak rates of incoming and outgoing flow;
- For each basin, the peak rates of inflow and outflow, maximum stage in ft. and maximum storage in cu.ft. or ac.ft.

You may want a different time step in which to do routing calculations. The model calculates a favorable time increment; changing it can greatly affect accuracy of results. Lengthening the time step might allow the model to proceed further with calculations for large members, but might cause rapid inflow hydrographs or short channel travel times to be missed entirely. Shortening the time step might make rapid hydrographs from small subwatersheds have more effect on peak flows downstream, but might cause peak flows from larger, slower members to be missed since the total calculation time could not proceed far enough. The time step should be less than the travel time through any member. As an alternative you could revise the way input data are entered, such as combining the smallest subwatersheds into ones of size more similar to others in the system, dividing the longest channels into ones more similar to other channels, etc.

You can print out a summary sheet of the runoff and routing calculations. If you want one, answer *Y* when the computer asks. This printout will be made automatically if you are running all six design storms. If a summary sheet is to be printed you will be asked to enter a title for the analysis. Your title can contain up to 50 letters or numbers, but not commas. You will also be asked whether the land use condition represented in the analysis is undeveloped, existing or developed. The title and land use condition are only printed on the output for that routing; they do not affect the results and are not stored on disk.

Check to make sure the results seem reasonable. This is a final confirmation of your input data and the model's selection of length and number of time steps.

You can run the data input module again, to work with the same or a new set of watershed and network data. If you answer *Y* you can enter new data or access data already saved on your data disk. If you answer *N* the program will end its operation and return the computer to the system mode.

Network Diagrams

A network diagram divides a watershed into its component subwatersheds, channels and reservoirs in order to identify where flows originate and their sequence of combinations downstream. Putting all this information together on a single diagram organizes data for entering into the computer.

The computer model demands that certain rules about data entry be followed. Although these rules make certain parts of the data a little more cumbersome than they would otherwise be, they allow the program to keep track of all the flow relationships in each network.

Three types of components need to be identified:

1. *Subwatersheds:* Subwatersheds generate the original volumes and timing of runoff that will be tracked down the rest of the drainage system.

2. *Channels:* Any conveyances, whether open or closed, and whether natural or man-made, which allow channel flow along their entire length. Even underground culverts flow under "open channel" conditions, as opposed to full flow which involves hydraulic pressure. A special kind of channel is a "dummy" channel, which transfers runoff from one member to another with no routing, that is, without any storage of water or passage of time during the transfer. Dummy channels are used only for connecting some subwatersheds into a network diagram. Dummy channels cannot be seen in a physical site plan; they exist only for the convenience of the computer model.

3. *Reservoirs:* Any segments of the drainage system from which the outflow is limited such that storm runoff may be stored in the segment. Such segments could be dry or wet detention basins, infiltration basins or undersized conveyances.

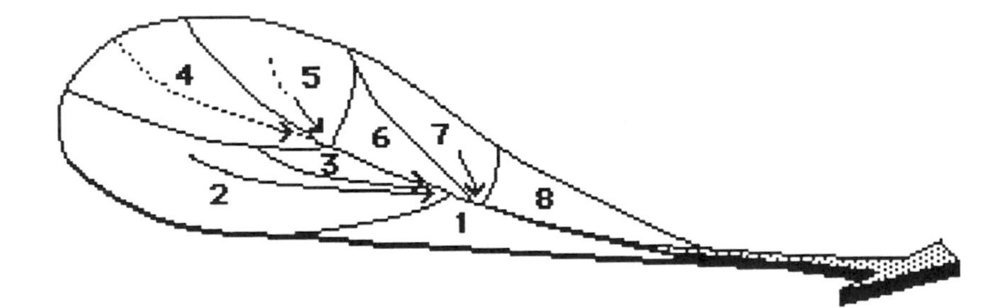

General steps for developing a network diagram are the following:

Outline the boundaries of the watershed on a map (see Appendix B, *Drainage Area Analysis,* if you need help with this step).

Identify the channel system within the watershed and locate all detention basins, infiltration basins, undersized conveyances and other reservoirs. The lengths of all channel segments in the system should be roughly similar, so that the computer can derive a meaningful hydrograph time increment for all of them. To equalize channel lengths, divide the longest segments into shorter ones more similar to other lengths in the network.

Divide the watershed into subwatersheds. The computer program will calculate a hydrograph for each subwatershed, for routing through the channel and reservoir system. In defining a subwatershed, all runoff from a subwatershed must leave the subwatershed at only one location, its outlet. The outlet from each subwatershed must be the upstream end of a channel, a dummy channel or a reservoir.

Give an indentifying number to each subwatershed, beginning with number 1. Then number all channels and reservoirs, again beginning with number 1, starting with all the headwater members that receive runoff directly from subwatersheds. The following rules must be followed:

All headwater segments which receive runoff directly from subwatersheds must be numbered first beginning with number 1. The identifying number of each of these segments must be the same as that of the subwatershed that drains into it. Only one subwatershed can flow into a receiving channel or reservoir; in other words, each subwatershed must flow into its own receiving channel or reservoir. Where necessary in the network diagram, insert "dummy" channels to connect subwatersheds into the drainage network. Since each dummy channel receives subwatershed runoff directly, each dummy must be given the same number as the subwatershed that drains into it.

Other segments in the network are numbered starting at the top of the watershed and working down through the watershed to the outlet. Lower-numbered segments must always flow into higher-numbered segments.

A drainage network analyzed in the model can have up to 50 subwatersheds. It can have up to 20 reservoirs, and a total of 100 channels and reservoirs. Each reservoir can have up to 10 "rating" values (stage-discharge-storage values). The model interpolates linearly between all entered values.

On the following pages are some example networks. The following symbols are used:

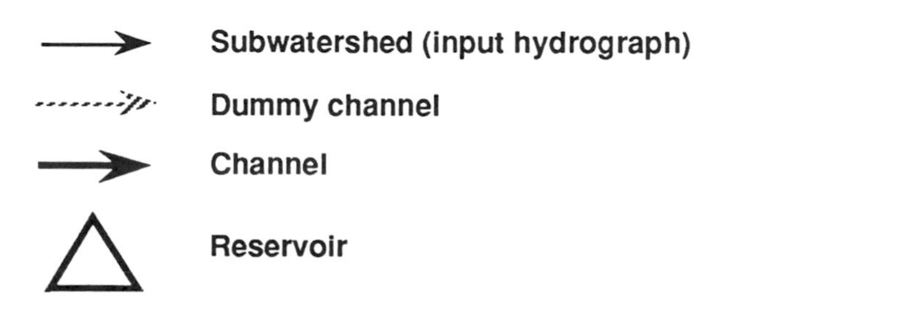

A watershed with one land use area and one detention basin:

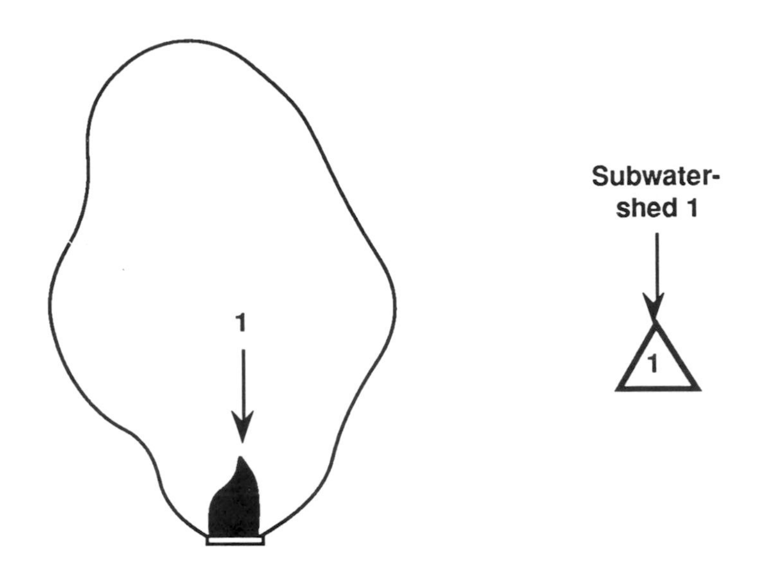

A watershed with three different land use areas:

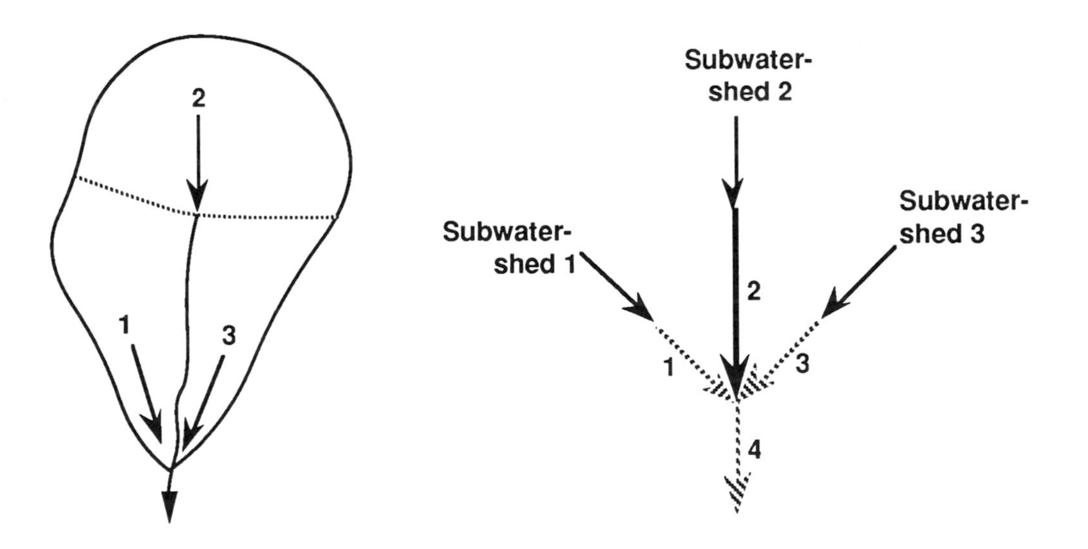

A watershed with two subwatersheds, one draining into an infiltration basin:

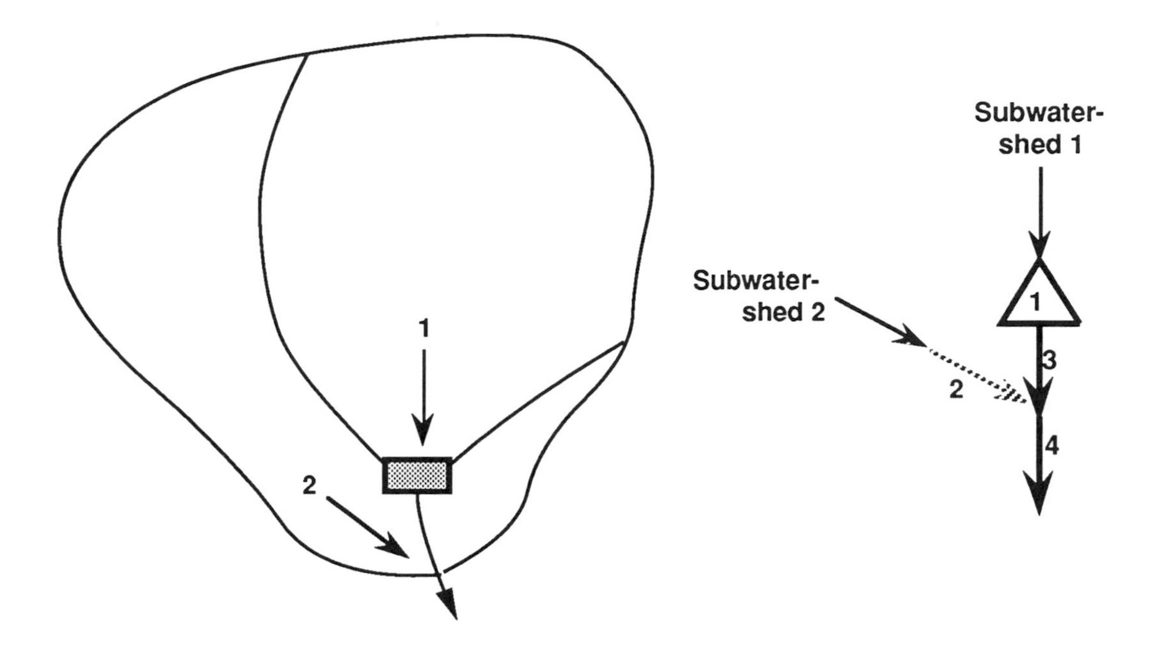

A watershed with two detention basins, each draining one subwatershed:

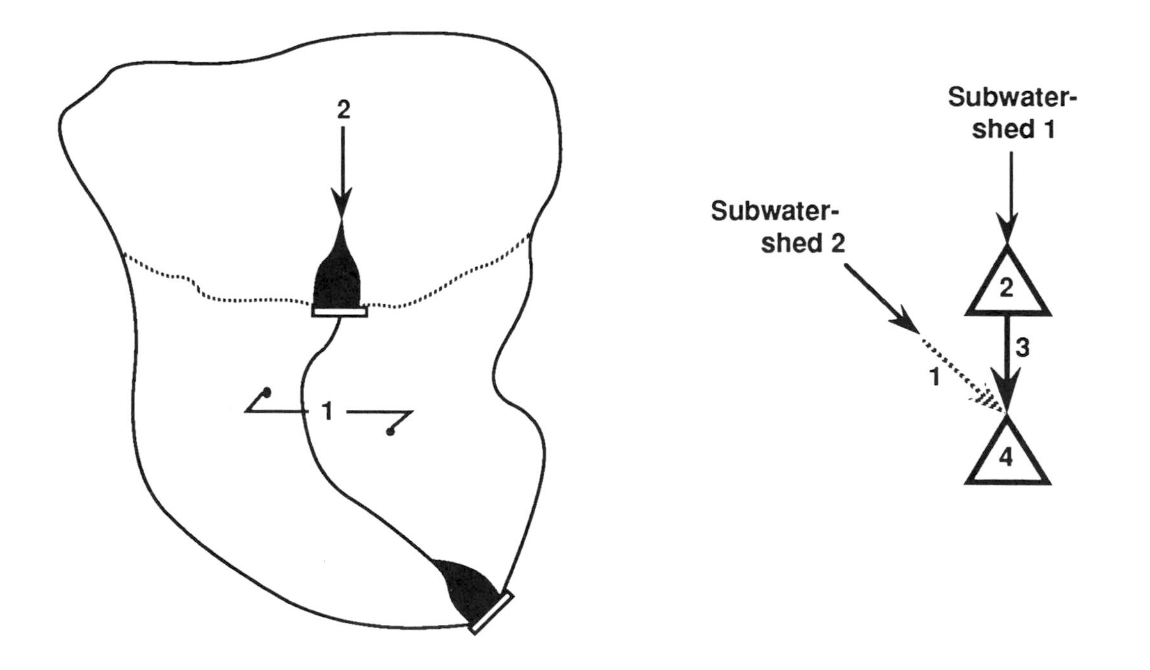

A watershed with nine land use areas and three detention basins:

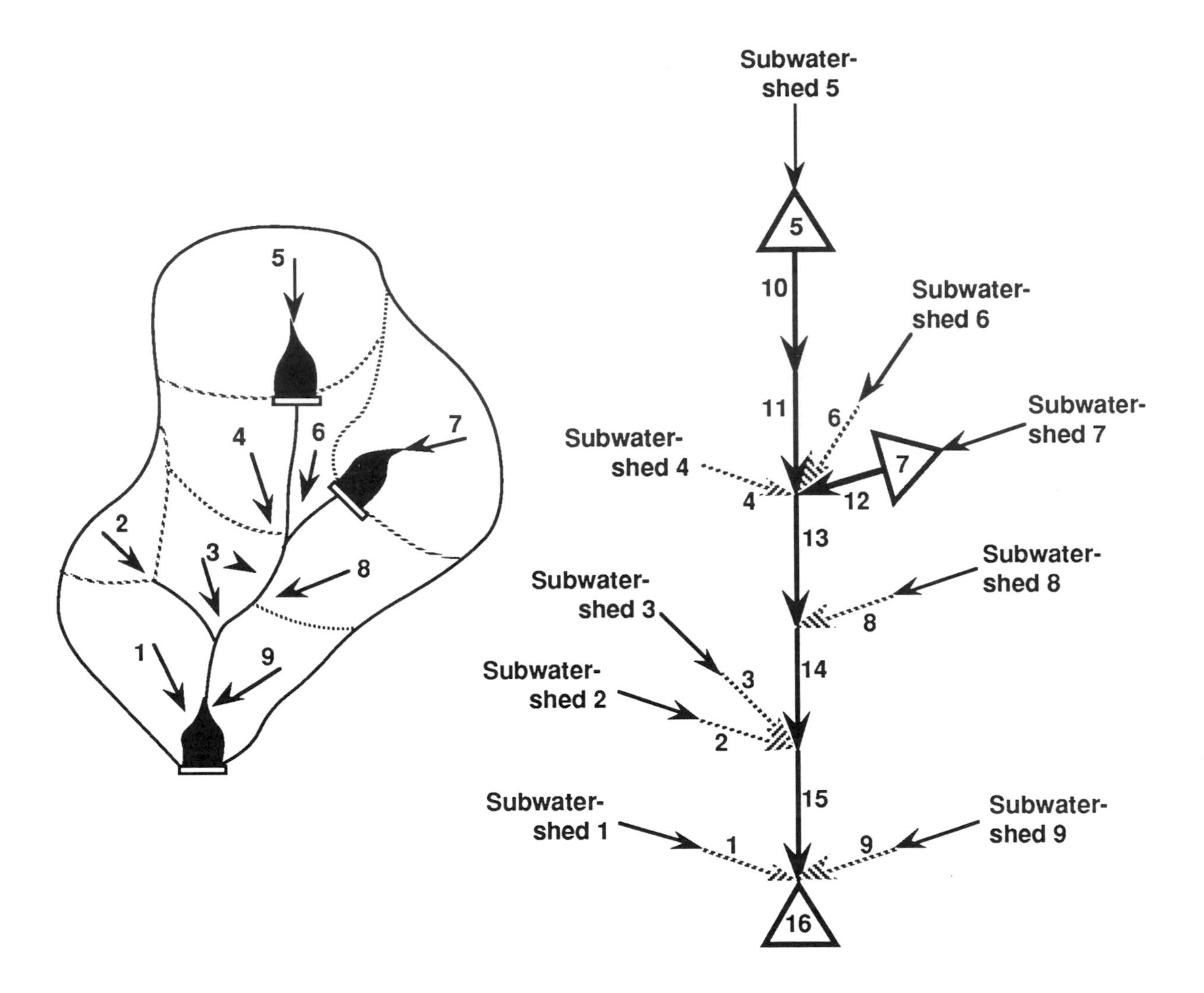

Possible Error Messages

Many errors in data input, etc. can be easily fixed when you know what the error messages mean. Following are some of the error messages your computer might give you when running the stormwater program. Most errors result from data input errors or just computer operation problems like having the correct disk in the correct drive. The program is written in Turbo Basic; if necessary, you might wish to refer to the Borland International's *Turbo Basic Owner's Handbook* for a complete list of errors that might be associated with the way this programming language operates.

File Not Found in Some Location — Did you give a correct file name ("job name") for the computer to look for? Did you enter subwatershed and other data correctly? One way this error is generated is a reversal of a subwatershed's 85% and 10% elevations, generating a negative slope that the computer cannot deal with. Check consistency of units of impervious area, whether acres or percent. Make sure channelized length is equal to or less than total length.

Illegal Function Call — Check all entered data for inconsistencies. Typical entry errors include channelized channel length greater than total channel length, impervious area greater than total area, and elevations entered in reverse order. Review the "data map" on the screen, where the model calculates useful indicators such as watershed slope and percent impervious.

Redo From Start — You are trying to enter data in a format different from that required by the model or the computer's operating system. Enter all required data by following the instructions in this manual or given on the computer screen.

Program Stops Running — Is the printer on if you want a printed copy?

You Anwered the Question but Nothing Happens — Did you press *Return*?

Error 7 — Out of memory. You computer's memory capacity was exceeded.

Error 11 — Division by zero. The data you entered requires the computer to divide by zero or to raise zero to a negative power. Check your input data.

Error 25 — Device fault: a hardware error has occurred. For example, is the printer correctly connected to the computer?

Error 27 — The printer is out of paper or needs to be turned on.

Error 53 — File not found. Check your default drive specifications in the STARTUP module. Is the right disk in the right drive?

Error 57 — Device input/output error: a hardware problem has occurred. Is the printer turned on?

Error 61 — Disk full. Try replacing your data disk with a new formatted disk or a high-density disk, or store data on your hard disk instead of a floppy.

Error 62 — Input past end. The model is trying to read more data than are available. Check your input data for inconsistencies.

Discussion Questions

The great speed of a computer at doing hydrologic calculations makes it possible to answer many "what if" questions quickly and precisely, after basic data for a given site have been entered. The following questions assume that stormwater control basins have been proposed in developed subwatersheds, and a routing calculation has been done using one reasonable storm recurrence interval.

1. What is the effect on discharge after development of adding a larger number of detention basins in one or more of your watersheds? What is the effect on area and volume of detention basins, totaling all detention basins together?

2. What is the effect on discharge, basin area and basin volume of moving each detention basin to a different location in your drainage system?

3. What is the effect on discharge, basin area and basin volume of increasing the impervious area in each subwatershed? You could assume that the non-impervious area is reduced by half. And what is the effect of 100% impervious area? What are some examples of land uses that might involve 100% impervious cover?

4. What is the effect on discharge, basin area and basin volume of using a significantly different recurrence interval for your design storm? If you have done calculations using a 10 to 25 year storm, try a 100 year storm.

5. If your watershed includes several subwatersheds with different proportions of impervious cover, what is the effect on discharge, basin area and basin volume of exchanging the land uses among the subwatersheds? Since their location in the watershed changes, their effect on timing of hydrographs might change too.

6. What are the effects on discharge, basin area and basin volume of significantly changing the proportion of major channels that are channelized?

7. What are the effects of varying times of concentration on flow volume and peak rate of flow? Why do these results respond to time of concentration in different ways?

Appendix I

Constants and Conversions

Units of numbers can be converted using the principle that the numerator and the denominator of a fraction can be multiplied by the same number (in essence multiplying the fraction by 1) without changing the value of the fraction.

To convert gallons per minute to cubic feet per day, we must first identify fractions that contains both the units of time (minutes and days) and the units of volume (gallons and cubic feet) and that, when they are used as multipliers, do not change the numerical value. Relative to time, there are 1,440 minutes in a day. Therefore, if any quantity of time is multiplied by 1,440 min./day, the result will be in different units, but its numerical value or meaning will be unchanged. Relative to volume, there are 7.48 gallons in a cubic foot. Therefore, to convert gallons per minute to cubic feet per day, we multiply by these "unit" fractions, cancel the units of measurement that appear in both the numerator and the denominator, and gather together the units that remain.

Here is an example, converting gallons per minute to cubic feet per day:

$$\frac{\text{gallons}}{\text{minute}} = \frac{\text{gallons}}{\text{minute}} \text{ X } \frac{1{,}440 \text{ min}}{\text{day}} \text{ X } \frac{\text{cubic feet}}{7.28 \text{ gal}} = 192.5 \text{ cu.ft./day}$$

which tells us us that 1 gal/min (gpm) = 192.5 cu.ft./day.

Length

1 foot
= 103.8 mm
1 cm
= 10 mm
1 meter (m)
= 1,000 mm
= 100 cm
= 3.28 ft.
= 39.37 in.
1 yard
= 3 feet
= 0.9144 m
1 mile
= 5,280 ft
= 1.609 km
1 km
= 3,280.8 ft
= 1,000 m
= 0.6214 miles (mi)
1 degree of latitude
= 69.057 miles (mi)

Area

1 sq.in.
= 6.452 sq.cm
1 sq.ft.
= 929.0 sq.cm.
1 acre
= 43,560 sq.ft.
= 4,840 sq.yd.
= 4046.86 sq. meters
1 sq.mi.
= 640 ac.
= 2.59 sq.km.?
1 sq.meter
= 10.764 sq.ft.
1 sq.ft.
= 144 sq.in.
= 929.03 sq.cm.

Volume

1 cubic centimeter (cc)
= 0.061 cu.in.
= 0.0338 U.S. fluid ounces (fl.oz.)
1 liter
= 1,000 cubic centimeters (cc)
= 61.1255 cu.in.
= 33.815 U.S. fl.oz.
= 1.057 U.S. quart
= 0.2642 U.S. gallon
= 0.03532 cu.ft.

1 cubic meter
= 1,000 liters
= 35.315 cu.ft.
= 264.17 U.S. gal.
1 milliliter (ml)
= 1 cu.cm. (cc)
= 0.0610
cu.in.
= 0.0338 U.S. fl.oz.
1 cu.ft.
= 1,728 cu.in.
= 7.48 U.S. gal.
29.92 U.S. quart
= 28316.8 cu.cm. (cc)
= 28.3161 liters
1 ac.ft.
= 1.23 million liters
= 325,851 gallons
= 43,560 cu.ft.
= 1,230 cu.meters
= 0.504 sceond-foot-day (cfs-days)
1 second-foot-day
= 0.646 million gallons
= 1.98 ac.ft.
= 2.45 million liters
= 2,450 cu.m.
1 cfs/sq.mi.
= 13.58 in/yr
1 U.S. gallon
= 0.134 cu.ft.
= 231 cu.in.
= 128 fl.oz.
= 3785.41 cu.cm. (cc)
= 3.785 liters
1 million U.S. gal (mg)
= 134,000 cu.ft.
= 3.076 ac.ft.
1 square mile-inch
= 2,315,000 cu.ft.
= 53.3 ac.ft.
1 day second foot (dsf)
= 1 cfs for 1 day (cfs-day)
= 86,400 cu.ft.
= 1.98 ac.ft.
1 acre-inch
= 3,630 cu.ft.

1 hectare-meter (ha-m)
= 10,000 cu.m.
= 8.107 ac.ft.

Time
1 minute
= 60 seconds (sec)
1 hour (hr)
= 60 minutes (min)
1 day
= 24 hours (hr)
= 86,400 seconds (sec)
1 year (yr)
= 8,760 hours (hr)

Velocity
1 centimeter per second (cm/sec)
= 0.03281 foot per second (ft/sec, fps)
1 kilometer per hour (km/hr)
= 54.68 feet per minute (fpm)
1 foot per minute (fpm)
= 60 feet per second (fps)
1 mile per hour (mph)
= 44.7 cm/sec
= 88 fpm
= 1.467 fps
= 1.609 km/hr
= 0.4470 meter/sec
1 meter per second
= 3.6 km/hr
= 0.621371 mi/hr (mph)
= 0.9113 ft/sec (fps)
1 foot per second (fps)
= 0.6818 mph
= 60 ft/min (fpm)
= 0.3048 m/sec
= 1.097 km/hr

Flow rate
1 million gallons per day (mgd)
= 1.55 cfs
= 3.07 ac.ft./day
= 0.0438 cu.meters/sec?
1 cubic foot per second (cfs)
= 0.646 mgd cu.meters/hr
= 1.98 ac.ft./day
= 2447 cu.meters/day
= 0.0283 cu.m/sec
= 449 gpm
= 28.32 liters/sec
1 gallon per minute (gpm)
= 0.0631 l/sec
= 0.00442 ac.ft/day
= 0.00223 cfs
= 0.0000631 cu.m/sec

1 cfs/sq.mi. (csm)
= 0.03719 inch/day
= 0.656 cu.m./minute/sq.km.
= 0.011 cu.m./sec/sq.km.
1 cu.m/sec
= 22.8 mgd
= 3.05 million cu.ft./day
= 86,400 cu.m./day
= 1,000 l/sec
= 70.0 ac.ft./day
= 35.32 cfs
1 in/hr of runoff from one acre
= 1 cfs

Mass
1 slug
= 14,600 grams (g)
= 32.17 pounds (lb)
1 kilogram (kg)
= 1,000 grams (g)
= 2.205 lb

Density
1 g/cu.cm.
= 62.428 lb/cu.ft.
1 lb/cu.ft.
= 0.01602 g/cu.cm

Energy
1 Btu
= 778.26 ft-lb
1 calorie (cal)
= 1 gram-calorie (g-cal)
= 0.003969 Btu
1 langley (ly)
= 1 calorie per square centimeter
= 3.69 Btu/sq.ft.

Power
1 Btu/minute
= 175.4 watts
1 Btu/sq.ft./minute
= 0.2713 cal/sq.cm./minute
= 0.01893 watt/sq.cm.

Temperature
Degrees Fahrenheit (°F)
= (9/5)(degrees Celsius) + 32
Degrees Celsius (°C)
= (5/9)(degrees Fahrenheit - 32)
Degrees Kelvin (°K)
= (degrees Celsius) + 273.16

Pressure
1 millibar (mb)
= 1,000 dynes/sq.cm.
= 0.0145 pounds.sq.in. (psi)
= 0.00102 kg/sq.cm.
= 0.7501 mm Hg
1 standard millimeter of mercury (mm Hg)
= 1.33 mb
= 0.00136 kg/sq.cm.
= 0.0394 in Hg
1 inch of mercury
= 33.9 mb
= 0.492 psi
1 standard atmosphere (atm)
= 1,013.25 mb
= 760 mm Hg
= 14.7 psi
= 1.033 kg/sq.cm.
= 29.92 in. Hg
1 pound per square inch (psi)
= 2.036 in. Hg
= 68.95 mb
= 0.0703 kg/sq.cm.
= 51.72 mm Hg

Properties of water
Latent heat of fusion
= 79.7 cal/g at 0°C
Latent heat of vaporization
= 597.3 cal/g at 0°C
Latent heat of sublimation
= 677.0 ca./g at 0°C
Density
= 1.94 slugs/cu.ft.
Specific heat
= 1 cal/g/°C at 14.5°C
Specific weight
= 62.4 lb/cu.ft.
= 2,720,000 lb/ac.ft.
= 1000 kg/cu.m.
= 8.33 lb/U.S.gal.
= 10.00 lb/imp.gal.
= 1 gram/cc
1 ppm of constituent in water
= 2.7 lb constituent/ac.ft. of water

Other natural properties
Falling speed of raindrops
= 618 m/sec (p. 51 of Miller, 1977)
Falling speed of snowflakes
= <1 m/sec (p. 51 of Miller, 1977)

Index

Accuracy of storm runoff estimation, 53
also see Rainfall-runoff models
Aeration of pond, in extended detention, 126
Aesthetics, wet basin for, 164
After-development conditions
see Developed conditions
Agricultural areas, Curve Number, 40
Alternatives in hydrologic functions
see Functions, hydrologic
Area
Relation to rate of flow, 7
Relation to volume, 6, 154
Area of floor of dry basin, 156-158
Area of pond, 126, 164-165
Estimation based on volume, 127
Arid areas, 153
Arizona, Phoenix, 196-197
Average of watershed conditions, weighted by area, 108

Background flows
  Estimation, 56
  Importance as resource, 56
  Importance to infiltration, 56, 145
Barrel pipe, 96
Base flows
  Definition and importance, 10
  Effect of storm detention, 109
  Relation to volume of storm flow, 136
  also see Low flows, Flood flows
Basins
  For infiltration, 140, 227-230
  For storm detention, 107
  Grading, 211-214
  Location, 120
  also see Dry basins, Wet basins,
    Ephemeral basins, Underground basins,
    Ponds, Reservoirs
Bedrock, relation to infiltration, 147
Before-development conditions, 106, 209-210
  also see Natural conditions
Bibliography, 177-184
Blaney-Criddle method, 171
Buildings along drainage area boundary, 205
Calibration of storm runoff estimation, 53
California
  Fresno, 139
  San Diego, 29
Catchment area
  Relation to monthly runoff, 66
  In water harvesting, 153, 154-155
  also see Drainage area, Watershed
Channel routing, on computer software, 237
  also see Reservoir routing
Channels, 14, 79
  Computer output, 249
  Data in computer software operation,
    239, 244, 250-254
    also see Drainage networks
  In SCS time of concentration, 223
Colorado, Denver, 30, 173
Commercial sites, 166
    also see Shopping centers
Compound conveyances, 99
Computer software
  Calibration to local conditions, 53
  In storm detention, 116
  Operation, 235-256
    also see Data in operation
  Reservoir routing, 111
  Time of concentration, 24
Concentration of constituent
  Definition, 11
  Relation to rate of flow, 11
Concrete
  Pipe, 88
  Swale, 90-91
Conservation of water, 174
Constituents, types and sources, 12
  also see Water quality
Construction, sediment during, 216
  also see Sediment
Contour lines
  At proposed basins, 211-214
  Indicating drainage area boundaries, 202-206
  also see Topography
Control of stormwater
  see Functions, hydrologic
Converting units, 257-260
Conveyance, 79-104
  Definition and history, 14
  Design, summary of process, 100
  Importance, 14, 104
  Selection of storm detention outlet, 114
  Safety in, 232
Corrugated metal pipe, 89
Cover factor in "rational" method, 26-27
  Average weighted by area, 208
  also see Efficiency of monthly runoff
Culverts
  In compound conveyance, 99
  In storm detention, 114
  Inlet control, 92-93
  Outlet control, 94-95
  also see Pipes, Manning's equation
Curve Number, 34, 38-41
  Average weighted by area, 208
Cut, see Excavation
Dams, 96
  Elevations at, 213
  In grading plan, 212
Danger, see Risk, Liability, Safety, Uncertainty
Darcy's equation, 74
Data in computer software operation
  Editing, 246-247
  Needed to enter, 239, 241, 243-244, 248
Days per month, 67
Dead storage, in extended detention pond, 126

Depth
In dry basin ponding, 158
In extended detention pond, 126
In storm detention basin
Relation to dam, 213
Relation to pond area, 165
also see Stage
In swale, 81, 90-91
In wet infiltration basin
Effect on water quality, 139
Of monthly runoff, estimation, 66
Of runoff, relation to volume of runoff, 42
Of storm runoff, estimation, 34, 42
Relation to pressure, 92
Relation to area and volume, 6
also see Head, Stage
Design, see Urban design
Design storms
Comparison to water balance, 56, 78
Computer software usage, 239
Importance, 10
In dry basin design, 158
In extended detention design, 125
In infiltration design, 145-146
In rainfall-runoff models, 20
In wet basin design, 165
Multiple, 115
also see Compound conveyances
also see Duration, Intensity, Recurrence interval
Detention
Definition, history and importance, 15-16
Relation to conveyance, 15
also see Storm detention, Extended detention
Detention basins
see Basins, Reservoirs
Developed conditions, 135, 209-210
Developer's role in storm detention, 116, 120
Diameter of pipe, see Culverts, Pipes
Disposal of stormwater
see Conveyance, Detention
Distribution of rainfall intensity, 43-46
also see Rainfall distribution types, Intensity
Drainage area
Analysis, 201-210
Boundaries, 201-207
In rainfall-runoff models, 20, 22, 26
Relation to peak rate of runoff, 43
Relation to volume of runoff, 42
also see Catchment area, Watershed
Drainage networks, 243-245, 250-254
Diagrams, 250-254
Drawdown time, see Ponding time
Dry basins, 15, 18, 107-108, 156-163
In infiltration, 140
Detention, grading, 212
Duration of storm
In "rational" method, 28
also see Rainfall, Intensity of rainfall
Earthwork
Effect on watershed boundary, 209
Relation to infiltration basin, 147
also see Grading, Excavation, Fill
Efficiency
Of irrigation system, 172
Of monthly runoff, 66, 153
El, see Evaporation
Elevation of water surface
In storm detention, 107
In relation to dam, 213
also see Head, Stage, Depth
Emergency overflow
Elevation of, 213
In grading plan, 212
In infiltration basin, 145
In storm detention, 109
Emergency spillway, see Emergency overflow
Energy dissipator, 214
Ephemeral basins
Open, 227-230
Summary of design process, 229
Underground, 142-148
Erosion, protection at detention outlet, 109
Error messages, in software operation, 255
Estimation of storm runoff, 19-54
Choice of methods, 54
Importance, 13, 19
"Rational" method, 26-32
SCS method, 24, 33-50
also see Rainfall-runoff models
Estimation of water balance
see Water balance estimation
Et, see Evapotranspiration
Evaporation
Estimation, 73
In open ephemeral basin design, 228
In wet basin design, 165
also see Evapotranspiration

Evapotranspiration
  Estimation, 73
  In dry basin design, 157
  Potential, 171
  also see Evaporation
Excavation, relation to infiltration basin, 147
Exercises
  Comments upon, 25, 210
  Computer software operation, 256
  Conveyance, 101-103
  Evapotranspiration, 76, 77
  Extended detention, 130-131
  Infiltration estimation, 76, 77
  Infiltration design, 149-151
  Monthly runoff estimation, 68-69, 77
    Including snow, 71-72, 77
  "Rational" method, 32, 51-52
  SCS method, 49-52, 225-226
  Sites, 187-200
  Storm detention, 118-119
  Time of concentration, 25
    SCS method, 225-226
  Water harvesting for dry basin, 161-163
  Water harvesting for wet basin, 168-170
Existing conditions
  Conveyances, 104
  also see Before-development conditions, Natural conditions
Extended detention, 121-132
  Combination with storm detention, 128
  Definition, history and importance, 16
  Effect on water quality, 16
  Summary of design process, 129
  also see Settling basins
Facilities for stormwater management
  Conveyance, 14
  Extended detention, 16, 128, 132
  Infiltration, 17, 140, 152
  Storm detention, 15, 120
  Water harvesting, 18, 156, 164, 171-172
  also see Dry basin, Wet basin, Ephemeral basin, Underground basin
Fill, relation to infiltration basin, 147
First flush
  Definition and importance, 11
  In infiltration design, 146
  also see Storm runoff, Water quality
Flood flows
  Control, see Storm detention
  Damage from, 105
  Definition and importance, 10
  also see Peak flows, Storm flows
Floor of dry basin, 156-158
Florida
  Longboat Key, 141
  Orlando, 144, 159
  Sarasota, 159
  Use of "littoral" zone, 126
  Use of "retention", 139
  Use of ephemeral basins, 142
  West Palm Beach, 141
Flow path
  see Length of travel, Hydraulic length
Flow rate, see Rate of flow, Volume of flow
Freeboard at dam, 213
Frequency of storm
  see Rainfall intensity, Recurrence interval
Functions, hydrologic
  Alternatives for, 5, 176
  also see Conveyance, Storm detention, Extended detention, Infiltration, Water harvesting
Georgia, Atlanta, 20-21, 188-189
Grading of basins, 211-214
Grading of site
  Relation to infiltration basin, 147
Graphical peak discharge method, 47
Groundwater
  Effect of infiltration, 152, 138-139
  Effect of storm detention, 109
Groundwater table
  Relation to infiltration basin, 140, 158
Harvest area, 154-155
Head
  In conveyance, 92-99
  also see Depth, Stage
High points on watershed boundary, 204-206
Hydraulic conductivity, 74-75
  also see Soil infiltration rate
Hydraulic head, see Head, Depth, Stage
Hydraulic length, 24, 219
  also see Time of concentration
Hydraulic radius, 83, 86-87
Hydrograph
  Definition, 9
  Generated by computer software, 237
  Long-term, 10
  Storm event, 9

Hydrologic function, see Functions, hydrologic
Hydrologic Soil Group, 37
  Effect on Curve Number, 38-40
Illinois
  Chicago, 122
  Drowning case, 233
  Riverside, 14
Impact, see Developed conditions
Impervious area
  Effect on Curve Number, 38-41
Industrial sites, 108, 141
  also see Office sites
Infiltration, 133-152
  Definition, history and importance, 17
  Effect on groundwater, 138-139
  Effect on water quality, 17
  Importance of water balance, 56
  In software operation, 245
  Sediment in basins, 215-218
  Summary of design process, 148
  also see Basins, Reservoirs,
    Soil infiltration rate
Inlet control at culvert, 92-93
Intensity of rainfall
  In "rational" method, 26, 28
  In SCS method, 43-44
  Relation to storm duration, 28
  also see Rainfall intensity
Invert elevation, 80, 107
Iowa, Waterloo, 166
Irrigation
  System efficiency, 172
  Water harvesting for, 18, 153, 171-173
  Water requirements, 171
Issues of stormwater management
  see Objectives of stormwater management
Lake evaporation, see Evaporation
Land surface
  Relation to flow velocity, 23
  also see Cover factor, Curve Number
Landfill site, example, 141
Landscape design, see Urban design
Landscape irrigation, see Irrigation
Length of travel
  Relation to time of concentration, 24
  Relation to travel time, 22
  also see Hydraulic length
Liability, 5, 231-234
  also see Risk, Safety
Littoral zone, 126
Live storage, in extended detention pond, 126
Long-term water balance
  Importance, 10
  Importance of estimation, 13
  In extended detention design, 125
  also see Water balance
Louisiana, drowning case, 234
Low flows
  Definition and importance, 10
  also see Base flows
Maintenance of infiltration basins, 218
Manning's equation, 83-91
  For channel travel time, 223
  also see Roughness
Maryland
  Glen Burnie, 144
  Use of infiltration, 146
Mass of constituent
  Relation to concentration, 11
Massachusetts, Canton, 108
Mathematics, 5, 6
Maximum probable storm, 21
Member numbers in drainage network, 251
Metal pipe, 89
Metals
  In pond sediment, 122
  In stormwater, 12
Methods, selection of, 175
  also see Rainfall-runoff models
Metric system, 257-260
Minnesota, Minneapolis, 30
Minnesota method of infiltration design,
  145, 228
Modeling of natural processes
  see Estimation of storm runoff,
    Estimation of water balance,
    Rainfall-runoff models
Monthly runoff
  Conversion to daily runoff, 67
  Estimation, 66
  In dry basin design, 157
  In extended detention design, 125
Multiple design storms, see Design storms
Multiple use developments, example, 166
Multiple use
  Of dry basins, 156, 158-159
  Of ponds, 128
Muskingum routing coefficient, 239

n, see Roughness
Natural areas, Curve Number, 40
Natural conditions, 134
  also see Before-development conditions
Natural cover in swale, 91
Natural wetlands in extended detention, 128
Networks, drainage, 243-245, 250-254
  Diagrams, 250-254
New Jersey
  Atlantic City, 190-191
  East Brunswick, 232
New Mexico
  Albuquerque, 194-195
  Rio Rancho, 108
New York, 26
  Albany, 196-197
  Ithaca, 192-193
  Long Island, 138-139, 159, 166
  Rochester, 192-193
  Syracuse, 141
North Carolina, Charlotte, 166
Numbers identifying network members, 251
Nutrients in stormwater, 12
Objectives of stormwater management,
    4, 10, 13, 176
  Conveyance, 79-81
  Infiltration, 133-139, 146
  Storm detention, 106
Office sites, examples, 159, 166
  also see Industrial sites
Ohio, Columbus, 108, 166
Olmsted, Frederick Law, 14
Open basins, ephemeral, 227-230
Open-channel flow
  Relation to time of concentration, 223
Outflow from basin
  Extended detention pond, 128
  Storm detention basin, 110-115
  Wet basin, 165
  also see Outflow, Emergency overflow,
    Rate of flow
Outlet control at culvert, 94-95
Outlet from basin, see Outflow
Outlet from drainage area, 202-203
Overlap of catchment and harvest areas, 155
Overlap of catchment area and basin, 228
Overload of stormwater facilities, 21
  also see Secondary system,
    Emergency overflow
Overflow
  see Overload, Secondary system,
    Emergency overflow
Parking lots, effect on pollutant loads, 139
Pavement, porous, 143
Peak flow
  Definition and importance, 10
  Effect of infiltration, 137
  Effect of water harvesting, 173
  Estimation in SCS method, 43
  In conveyance design, 80
  In storm detention design, 106, 120
  also see Flood flow, Storm flow,
    Rate of flow, Volume of flow
Pennsylvania
  Philadelphia, 190-191
  State College, 232
Permeability of soil, see Soil infiltration rate
Pipes, 14, 79, 85, 86-87
  Used as weir, 96, 98
  also see Culverts, Manning's equation
Plant water requirements, 171
Plants, wetland, 128
Pollutants, see Constituents
Ponding time in dry basin, 158
Ponds
  In extended detention, 16
  Relation to peak flow estimation, 43, 47
  Safety in, 233
  Water harvesting to supply water for, 153
  also see Volume of pond, Area of pond,
    Wet basins, Wetlands
Porous pavement, 143
Precipitation
  Relation to monthly runoff, 66
  also see Rainfall
Precipitation, direct
  In open ephemeral basin design, 228
  In water harvesting, 155
  In wet basin design, 165
Primary drainage system, 21
Public, role in storm detention, 116, 120
Pumps for irrigation water, 172
Purposes of stormwater management
  see Objectives of stormwater management
Q, see Rate of flow
Qp, see Peak flow, Storm flow, Flood flow
Qualitative choices in stormwater management
  see Functions, hydrologic

Quality, see Water quality
Qvol, see Volume of flow
Rainfall
  Distribution types
    In computer software operation, 248
    Relation to storm detention volume, 113
    also see Distribution of rainfall intensity
  Duration, in "rational" method, 28
  Intensity-duration-frequency relationship, 28-30
  In computer software operation, 239
  In rainfall-runoff models, 20
  In SCS method, 35-36
  Sources of data, 35
  also see Intensity of rainfall, Precipitation
Rainfall-runoff models, 20
  also see "Rational" method, SCS method, Monthly runoff, Water balance
Rate of flow
  Definition, 7
  Estimation by SCS method, 34
  Estimation by "Rational" method, 26-32
  Of constituent, 11
  Relation to constituent concentration, 11
  Relation to volume and time, 7
  also see Peak flow, Volume of flow
Rating curve
  see Stage-storage-discharge relationship
"Rational" method, 26-32
  Different versions, 28
  Summary of process, 31
Recharge
  On Long Island, New York, 138-139
  also see Infiltration
Recreation
  Wet basin for, 164
  also see Multiple use
Recurrence interval
  Definition, 20
  In computer software operation, 248
  also see Design storm
Recycling of water
  By infiltration, 138
  also see Infiltration, Water harvesting
References, 177-184
Regimes, hydraulic
  see Dry basin, Wet basin, Ephemeral basin
Regional systems of storm detention, 116, 120
Reservoir routing, 111
  For infiltration, 145
  By computer software, 237
  For open ephemeral basin, 228
  also see Channel routing
Reservoirs
  Data needed for software operation, 244, 250-254
    also see Drainage networks
  Software output, 249
Residence time
  Definition, 123-124
  Relation to pond volume, 124
Residential sites, examples, 108, 141, 159
Responsibility for design, see Liability
"Retention"
  Practices in Florida, 139
  also see Detention, Infiltration, Wet basins
Riprap, 90-91, 214
  also see Stone, crushed
Risk
  In relation to design storm selection, 21
  also see Liability, Safety, Uncertainty
Roads on drainage area boundary, 206-207
Roughness, in Manning equation, 83, 84, 86-91
  In sheet flow, 221
Routing
  see Channel routing, Reservoir routing
Runoff
  see Rate of flow, Volume of flow, Depth of runoff, Monthly runoff
Runoff coefficient
  see Cover factor, Efficiency of runoff
Runoff efficiency
  see Cover factor, Efficiency of runoff
Runoff irrigation, 171
Saddle points on watershed boundary, 204
Safety, 231-234
Safety factor, use of, 175
Safety factor on infiltration rate
  In dry basin design, 156-157
  In open ephemeral basin design, 228
  In wet basin design, 164-165
  also see Soil infiltration rate
SCS method
  Comparison to "rational" method, 33
  On computer software, 235-242, 248-249
  Summary of process, 48
  Time of concentration, 24, 219-226
    Summary of process, 225

Secondary drainage system, 21
  also see Emergency overflow
Secondary overflow, in storm detention, 109
  also see Emergency overflow
Sediment
  Control during construction, 128
  In basins, 215-218
  In ponds, 121-122, 128
  Removal by extended detention, 16
  Types, 216
  also see Constituents, Water quality, Settlement of sediment, Settling basins
Selection of calculation methods, 175
  also see Rainfall-runoff models
Settlement of sediment in ponds, 121-123, 128
Settling basins, 217
  In combination with infiltration, 139
  also see Extended detention
Shallow concentrated flow
  In time of concentration, 222
Sheet flow
  In time of concentration, 220-221
Shopping centers
  Effect on pollutant loads, 139
  Examples, 144
  also see Commercial sites
Site design
  see Urban design, Facilities for stormwater management
Size of pipe or culvert, see Pipe, Culvert
Slope, in conveyance, 83, 86-91
Snow
  In monthly runoff estimation, 70
  In storm runoff estimation, 54
Social demands upon stormwater
  see Objectives of stormwater management
Soil
  Relation to cover factor, 26-27
  Relation to Hydrologic Soil Group, 37
  also see Hydrologic Soil Group, Cover factor
Soil Conservation Service, see SCS method
Soil drainage
  Effect on Hydrologic Soil Group, 37
Soil infiltration rate, 74-75
  In dry basin design, 156-157
  In dry basin ponding time, 158
  In infiltration basin design, 148
  In open ephemeral basin design, 228
  In water balance estimation, 74-75
  In wet basin design, 164-165
Soil moisture, relation to Curve Number, 239
Soil permeability, see Soil infiltration rate
South Carolina, Charleston, 188-189
Southwestern United States, 153
Spillway, see Emergency overflow
Stage
  In storm detention basin, 110-115
  also see Head, Depth
Stage-storage-discharge relationship, 110
  In computer software operation, 244
Stone, crushed, 82, 143
  Hydraulic conductivity, 75
  also see Riprap
Storage
  In open ephemeral basin, 228
  In storm detention, 107, 110-115
  In subsurface, 134-137
  Of irrigation water, 172
Storm detention, 105-120
  Basin grading, 211-214
  Basin safety, 232-233
  Definition and importance, 15
  Ponding depth, relation to dam, 213
  Summary of process, 117
  Combined with extended detention, 128
Storm flow, see Peak flow, Flood flow
Storm runoff
  see Estimation of storm runoff, First flush
Stream flow
  see Base flow, Flood flow, Hydrograph
Subsurface
  Relation to infiltration, 147
  Hydrology, 134
  Storage in, 137
Surface stormwater systems
  see Conveyance, Detention
Swales, 14, 79, 80, 85, 90-91
Swamps, relation to peak rate of runoff, 43, 47
T, see Time
Technical Release 55, see SCS method
Texas
  El Paso, 20-21, 198-199
  Houston, 198-199
Time
  Relation to velocity, 7
  Relation to volume and rate of flow, 7
  Relation to rate of flow of constituent, 11
  also see Residence time

Time of concentration
Definition, 24
In computer software operation, 24, 248
SCS method, 219-226
Summary of process, 225
also see Time of travel
Time of ponding in dry basin, 158
Time of travel
Definition, 22
In open-channel flow, 223
In shallow concentrated flow, 222
In sheet flow, 220-221
also see Time of concentration
Time step in computer software operation, 249
Timing of floods and storm detention, 116
Topography
see Contour lines, Saddle points, Drainage areas
Toxic substances, see Metals, Nutrients, Sediment, Water quality
TR-55, see SCS method
Trap efficiency of reservoirs, 123
also see Settlement of sediment
Trapezoidal swale, 91
Travel time, see Time of travel
Treatment of stormwater, 16
also see Water quality
Triangular swale, conveyance in, 90
Tributaries, effect on flood flows, 116
Turf, in swale, 90-91
Types of stormwater management
see Functions, hydrologic
U.S. Soil Conservation Service, see SCS
Uncertainty, dealing with, 175
also see Risk
Underground basins, 142-143
Units, 257-260
Urban areas, Curve Number, 38-41
Urban design
In conveyance, 104
In extended detention, 132
In infiltration, 152
In storm detention, 107-108
In storm detention, 120
In water harvesting, 174
Use of water, see Water harvesting
Vegetation in dry basins
Protection from saturation, 156-157
Relation to sediment, 217
Velocity
Definition, 7
Estimation in conveyance, 85-91, 93, 95
In shallow concentrated flow, 222
Relation to area and rate of flow, 7
Relation to land surface, 23
Relation to swale erosion, 81, 82
Relation to time of concentration, 24
Relation to travel time, 22
Void space in crushed stone, 143
Volume of basin
Dry basin, 158
Extended detention pond, 121-126
Relation to area, 127
Relation to residence time, 12-125
also see Basins, Ponds, Residence time, Storage
Volume of flow
Effect of development, 136
Effect of extended detention, 132
Effect of infiltration, 137
In storm detention design, 120
Relation to constituent concentration, 11
Relation to depth and area, 6
Relation to inflow and outflow, 8
Relation to rate of flow and time, 8
Volume of monthly runoff, estimation, 66
Volume of storm runoff, estimation, 34, 42
Washington, Seattle, 20-21, 29
Water balance, 10
Basic equation, 57
Comparison to design-storm approach, 56
Data, 58-65
Estimation, 13, 55-78
Importance, 78
In dry basin design, 156-157
In infiltration design, 145
In open ephemeral basin design, 228
In wet basin design, 164-167
Of development site, 230
also see Long-term water balance
Water harvesting, 153-174
Definition, history and importance, 18
Dry basin design, summary of process, 160
Wet basin design, summary of process, 167
Water quality, 122, 123, 174
Control, see Functions, hydrologic
Effect of conveyance, 14, 81
Effect of extended detention, 132

Effect of infiltration, 137, 139
Effect of storm detention, 109
Importance, 11
Relation to irrigation, 171
also see Sediment, Constituents,
Concentration of constituent
Water supply by water harvesting, 153-155
also see Ponds, Irrigation
Water surface elevation, relation to dam, 213
Water table
see Groundwater, Groundwater table
Watershed
Data needed for software operation,
240-242, 250-254
Data on computer output, 249
also see Catchment area, Drainage area
Weirs, 96-99
In storm detention, 114
Wet basins, 15, 16, 18, 107-108
Elevations in, 213
Importance of water balance, 56
In extended detention, 121-128
In infiltration, 140
Water harvesting for, 164-170
Wet detention, see Extended detention
Wetland plants, in extended detention, 16, 128
Wetlands
Relation to peak flow estimation, 43, 47
Wet basin for, 164
Wildlife, wet basin for, 164

DEMCO